高等工科学校适用教材

ASP. NET MVC
实训教程

赵鲁涛　李　晔　汪兆洋　何森雨　编
王　策　杜云飞　薛美美

机械工业出版社

本书介绍了较为先进的 ASP. NET MVC 4 框架，对 MVC 4 进行了深入浅出的讲解，并通过实例实训的方式，让读者将理论联系实践，在动手操作中掌握核心知识；加入了 JQuery、MsChart、Flash 等前端应用技术，并将这些技术与 MVC 紧密结合，使知识和技能不是只停留在表层上，而是上升到深层应用上，提高了教材的整体层次，加强了学生对实际项目中实用技能的学习；充分发挥编者在实际项目中的经验，以多个实际工程项目为实例，从软件工程的角度出发，将整个工程从需求到最后完成的整个过程的关键技术完全复现在书中，不仅从技术上做到深化，而且从实际工程方面给学生更为清晰的范例，使学生能真实地感受到实际项目中 MVC 的具体应用。

图书在版编目（CIP）数据

ASP. NET MVC 实训教程/赵鲁涛等编. —北京：机械工业出版社，2015.6（2021.1 重印）
高等工科学校适用教材
ISBN 978-7-111-49853-7

Ⅰ.①A… Ⅱ.①赵… Ⅲ.①网页制作工具—程序设计—高等学校—教材 Ⅳ.①TP393.092

中国版本图书馆 CIP 数据核字（2015）第 067304 号

机械工业出版社（北京市百万庄大街 22 号　邮政编码 100037）
策划编辑：郑　玫　责任编辑：郑　玫　罗子超
版式设计：赵颖喆　责任校对：王　欣
封面设计：鞠　杨　责任印制：常天培
北京虎彩文化传播有限公司印刷
2021 年 1 月第 1 版第 4 次印刷
184mm×260mm・19.5 印张・484 千字
标准书号：ISBN 978-7-111-49853-7
定价：45.00 元

电话服务　　　　　　　　网络服务
客服电话：010-88361066　机　工　官　网：www.cmpbook.com
　　　　　010-88379833　机　工　官　博：weibo.com/cmp1952
　　　　　010-68326294　金　书　网：www.golden-book.com
封底无防伪标均为盗版　　机工教育服务网：www.cmpedu.com

序

随着互联网的发展，Web 开发技术得到了迅猛发展，软件行业对 Web 应用软件开发人员的要求也越来越高。因此，如何快速掌握目前开发所使用的主流技术和产品，并应用于现实生活和实际工作中，成为迫切需要解决的问题。为了适应这种需求，各高等院校开设了计算机相关实训课程。计算机相关专业的学生在学习计算机技术的过程中，需要通过生产实习，才能初步接触社会和认识社会，经受实际工作的基本训练，学会收集和整理信息资料，培养理论联系实践的能力，综合运用所学的知识，分析在科研或工程中实际遇到的问题，掌握解决实际问题的基本思路和方法。

ASP.NET MVC 是现阶段主流的开发框架，而国内目前对于 ASP.NET MVC 应用讲解的教材较少，大多数已有教材仅从架构角度进行描述，对于学生学习和教师讲授都不方便。有些教材没有将学生的学习特点考虑到其中，更多的是对有编程基础的程序员的一种晋级培训，无法为初学者提供一个较为完善的学习规范。

鉴于以上原因，编者编写了本书。本书主要参考北京科技大学计算机科学与技术系多名优秀教师的讲义以及反映较好的国内外相关书籍。编者主持和参与了多个教学科研项目，加上十几年的教学实践，积累了不少实际经验，对这门课程有了更深一层的领悟。本书的主要特点是：

（1）介绍了较为先进的 ASP.NET MVC 4 框架，对 MVC 4 进行了详解，另外，通过实例实训的方式，理论联系实践，让读者在动手操作中掌握核心知识。

（2）加入了 JQuery、MsChart、Flash 等前端应用技术，并将这些技术与 MVC 紧密结合。

（3）充分利用编者在实际项目中的经验，以多个实际工程项目为实例，从软件工程的角度出发，将整个工程从需求到最后完成的整个过程的关键技术完全复现在书中。

"学以致用，以用致学"是编者的教学理念，"学会学习"是学生在学习过程中的更高级要求，本书正是在这样的背景下，对 MVC 技术进行实例讲解，满足实践类课程的教学要求，让学生不断学习和掌握新知识。随着 MVC 技术的发展和成熟，本书亦将会不断更新和完善。

编 者

前　言

．NET 作为现在主流的开发平台，其 Web 开发一直采用 WebForm 模式，开发人员在体验着 WebForm 模式带来便利的同时，也体会到了视图与业务耦合造成的代码混乱。ASP．NET MVC 4 是微软公司官方提供的以 MVC 模式为基础的 ASP．NET Web 应用程序框架，其技术本身有着灵活、开发周期短、可重用性高等优点。

本书读者对象

本书由浅入深地对 ASP．NET MVC 4 进行了全面讲解，非常适合没有接触过编程的初学者学习使用，也适合作为计算机专业相关课程的教材使用。本书适合下列类型的读者使用：

（1）刚接触 MVC 4 开发，对 MVC 4 还不甚了解的初学者。

（2）对于 MVC 4 有一定的了解，但是并不娴熟，而且没有太多 MVC 4 开发经验的读者。

（3）希望找一本可以用于教授他人 MVC 4 开发方法的读者。

本书内容

本书从教学实际需求出发，合理安排知识结构，由浅入深、循序渐进地介绍较为先进的 ASP．NET MVC 4 框架。全书共分为 10 章，主要内容如下。

第 1 章 MVC 4 简介，主要介绍一般页面制作时需要掌握的知识、内容、思想、结构，并从 HTML、JavaScript、ASP．NET 的应用和 MVC 的原理出发，逐步深入讲解有关知识。通过本章的学习，读者能够对 MVC 4 有个大致的认识。

第 2 章模型，介绍 MVC 框架中模型（Model）层的执行机制，并介绍在 Visual Studio 2012 中如何利用 ADO．NET 实体模型来构造实体数据模型。在此基础上，介绍自定义数据模型和数据检验的相关内容。

第 3 章控制器，介绍控制器的功能与创建，Action 的处理流程，Action 在处理视图层和模型层之间的交互时的 4 种典型处理模式，以及 Action 的常见标签等内容。

第 4 章路由，介绍路由的基础知识、路由的解析、注册与管理，以及 ASP．NET MVC 执行生命周期等内容。

第 5 章视图，介绍如何写好页面的跳转逻辑以及运用一些原本在 WebForm 里面的 HTML 标签，使程序员可以顺利地从 WebForm 过渡到 MVC 的设计逻辑之中。

第 6 章 ActionResult 类，介绍如何利用 ASP．NET MVC 4 新增的区域（Area）机制构建较大的工程项目，以及如何将独立性较高的功能切割成多个 ASP．NET MVC 子网站，以降低网站之间的耦合性，降低在多人同时开发一个项目时发生冲突的概率。

第7章 JavaScript 与 JQuery 技术，介绍常用的前台技术：JavaScript 和 JQuery。从两种技术的基础开始介绍，逐步深入到高级应用，并汇集一些高级应用的范例使之更容易使用。

第8章 JQuery 高级应用，介绍利用 JQuery 编写的控件：zTree 和 JQGrid。本章主要内容包括这些树形控件和表格控件的调用方式、语句格式、作用以及使用它们可以达到何种效果等。

第9章 AJAX 技术，介绍从 AJAX 原理到 JQuery AJAX 中 3 种较为常用的实现方式，以及如何利用 AJAX 方式实现多属性查询功能。

第10章服务器（IIS）的配置与使用，选用 IIS 7 作为部署讲解的对象，通过一步步讲解 IIS 的安装、属性与配置，以及工程的发布，使读者清楚整个软件发布的流程。

本书图文并茂、通俗易懂、结构合理、内容丰富，在讲解每一部分时都给出相应的实例和表格进行说明，使读者更加容易理解并掌握相关知识。此外，本书侧重于技术层面，偏重应用，每一章后面都配有习题和综合应用，让读者在实践中巩固所学理论知识，快速提高操作技能。本书内容可按 80 学时讲授。本书第 1、2、3、5、10 章由赵鲁涛、王策、薛美美编写，第 4、6、9 章由李晔、何森雨、杜云飞编写，第 7、8 章由汪兆洋、何森雨、赵鲁涛、李晔编写。赵鲁涛负责本书的构思和统稿。

本书的编写得到了"十二五"期间北京科技大学教材建设经费资助和北京高等学校青年英才计划项目（YETP0386）的资助。在本书写作过程中，还得到了北京科技大学信息与计算科学系有关领导和同志的热情帮助和大力支持，在此表示衷心的感谢！同时，还要感谢北京科技大学李安贵教授和中国科学院软件研究所吴文玲研究员对本书提出的宝贵建议。

本书参考了中国建站之家（www.jz123.cn/）、.Net 源码服务专家（www.51aspx.com/Search/MVC 4）、jQuery（jquery.com）、W3School（www.w3school.com.cn）、zTree—jQuery 树插件（www.ztree.me/v3/main.php#_zTreeInfo）、jqGrid Home（http://www.jqgrid.com/）等网站的文章，在此对作者们一并表示感谢。

由于计算机技术发展日新月异，而 MVC 正是其中正在发展还未完全成熟的新兴技术，同时也由于作者的水平所限，书中错漏之处在所难免，敬请广大读者批评指正。

编　者

目　　录

序
前言
第1章　MVC 4 简介 ……………………… 1
1.1　基础知识 ……………………………… 1
1.1.1　HTML ……………………………… 1
1.1.2　CSS ………………………………… 7
1.2　基于 WebForm 的 ASP.NET ………… 9
1.2.1　.NET ………………………………… 9
1.2.2　ASP.NET …………………………… 9
1.2.3　WebForm 的创建方法 …………… 11
1.2.4　ASP.NET 服务器控件 …………… 12
1.3　MVC 概念与原理 …………………… 16
1.3.1　MVC 的概念 ……………………… 16
1.3.2　MVC 的工作原理 ………………… 16
1.3.3　MVC 架构的优缺点 ……………… 17
1.4　初识 MVC 4 ………………………… 18
1.4.1　创建新项目 ……………………… 18
1.4.2　MVC 4 项目 ……………………… 21
习题 ………………………………………… 31
综合应用 …………………………………… 31

第2章　模型 …………………………… 34
2.1　模型层概述与执行机制 ……………… 34
2.2　实体数据模型 ………………………… 35
2.3　LINQ 语句与使用 …………………… 37
2.3.1　使用 LINQ 的好处 ……………… 37
2.3.2　LINQ to SQL 的预备知识 ……… 38
2.3.3　LINQ to SQL 的查询 …………… 40
2.3.4　LINQ to SQL 进行插入 ………… 43
2.3.5　LINQ to SQL 进行更新 ………… 43
2.3.6　LINQ to SQL 进行删除 ………… 44
2.4　模型的数据校验 ……………………… 44
2.4.1　非数据库类 DataAnnotation 启用
　　　　验证 ………………………………… 44
2.4.2　数据库类 DataAnnotation 启用
　　　　验证 ………………………………… 52

习题 ………………………………………… 54
综合应用 …………………………………… 54

第3章　控制器 ………………………… 60
3.1　控制器概述 …………………………… 60
3.2　控制器的创建 ………………………… 62
3.3　Action 的处理流程 ………………… 64
3.3.1　参数获取 ………………………… 65
3.3.2　参数预处理 ……………………… 66
3.3.3　与模型层的交互 ………………… 66
3.3.4　结果预处理 ……………………… 66
3.3.5　视图返回 ………………………… 67
3.3.6　实例分析 ………………………… 67
3.4　典型的处理模式 ……………………… 68
3.4.1　单个视图调用多个函数 ………… 69
3.4.2　多个视图调用单个函数 ………… 71
3.4.3　多个视图调用多个函数 ………… 74
3.5　Action 的常见标签 ………………… 77
3.5.1　NonAction ………………………… 78
3.5.2　HttpGet 和 HttpPost ……………… 79
3.5.3　ChildActionOnly …………………… 80
习题 ………………………………………… 82
综合应用 …………………………………… 82

第4章　路由 …………………………… 87
4.1　路由的基础 …………………………… 87
4.1.1　网址路由的作用 ………………… 87
4.1.2　默认的 Route Table ……………… 88
4.2　路由解析 ……………………………… 89
4.2.1　非 MVC 控制器类路由解析 …… 90
4.2.2　带单个参数的 MVC 路由地址的
　　　　解析 ………………………………… 90
4.2.3　带多参数的 MVC 路由地址的
　　　　解析 ………………………………… 91
4.3　路由注册 ……………………………… 91
4.4　路由管理与匹配机制 ………………… 93

4.5　MVC 执行的生命周期 …………… 94
　　4.5.1　网址路由比对阶段 ………… 94
　　4.5.2　执行 Controller 的 Action
　　　　　阶段 ………………………… 95
　　4.5.3　执行 View 并返回结果页面 … 95
4.6　总结 ……………………………… 95
习题 …………………………………… 96
综合应用 ……………………………… 96

第 5 章　视图 ……………………… 97
5.1　视图概述 …………………………… 97
5.2　视图页 ……………………………… 98
　　5.2.1　视图页的创建 ………………… 98
　　5.2.2　视图页介绍 ………………… 101
5.3　从控制器层获取数据的方式 ……… 105
　　5.3.1　弱类型 ……………………… 105
　　5.3.2　强类型 ……………………… 109
　　5.3.3　Session 和 Cookies ………… 113
5.4　HtmlHelper 类 …………………… 114
　　5.4.1　ActionLink ………………… 115
　　5.4.2　BeginForm 和 EndForm …… 116
　　5.4.3　CheckBox ………………… 118
　　5.4.4　DropDownList ……………… 119
　　5.4.5　Hidden …………………… 122
　　5.4.6　Label ……………………… 123
　　5.4.7　ListBox …………………… 124
　　5.4.8　Password ………………… 125
　　5.4.9　RadioButton ……………… 126
　　5.4.10　TextArea ………………… 127
　　5.4.11　TextBox ………………… 129
5.5　布局页和视图布局页 …………… 130
　　5.5.1　布局页的创建 ……………… 130
　　5.5.2　视图布局页的创建 ………… 131
　　5.5.3　布局页和视图布局页介绍 …… 133
　　5.5.4　布局页的嵌套 ……………… 135
5.6　分部页 …………………………… 138
　　5.6.1　分部页的创建 ……………… 138
　　5.6.2　分部页介绍 ………………… 140
5.7　向控制器层传递数据的
　　　方式 ……………………………… 144
习题 …………………………………… 147
综合应用 ……………………………… 150

第 6 章　ActionResult 类 ………… 154
6.1　ActionResult 类概述 …………… 154
6.2　ViewResult ……………………… 156
6.3　PartialViewResult ……………… 158
6.4　ContentResult …………………… 158
6.5　EmptyResult …………………… 160
6.6　FileContentResult、FileStreamResult
　　　和 FilePathResult ……………… 160
6.7　JavaScriptResult ………………… 163
6.8　JsonResult ……………………… 165
6.9　RedirectResult …………………… 167
6.10　RedirectToRouteResult ………… 168
6.11　HttpUnauthorizedResult 和
　　　HttpNotFoundResult …………… 170
习题 …………………………………… 171
综合应用 ……………………………… 172

第 7 章　JavaScript 与 JQuery 技术 … 173
7.1　JavaScript ……………………… 173
　　7.1.1　JavaScript 简介 …………… 173
　　7.1.2　JavaScript 的语法 ………… 175
　　7.1.3　JavaScript 函数 …………… 178
7.2　JQuery 简介 ……………………… 179
　　7.2.1　选择器 ……………………… 180
　　7.2.2　JQuery 中的文件对象模型与
　　　　　方法 ………………………… 183
　　7.2.3　事件处理 …………………… 186
7.3　JavaScript 与 JQuery 应用
　　　实例 ……………………………… 187
　　7.3.1　iPhone 界面制作 …………… 187
　　7.3.2　使用 JQuery 给 table 动态添加、
　　　　　删除行 ……………………… 192
　　7.3.3　使用 JQuery 生成精美的 Tab
　　　　　按钮 ………………………… 194
　　7.3.4　使用 JQuery 完成相框效果 … 200
习题 …………………………………… 201
综合应用 ……………………………… 202

第 8 章　JQuery 高级应用 ………… 207
8.1　zTree 控件 ……………………… 207
8.2　zTree 的 API …………………… 210
　　8.2.1　API 综述 …………………… 210

8.2.2 常用 API 详解 …… 215
8.3 zTree 应用实例 …… 222
　8.3.1 zTree 基本功能 …… 222
　8.3.2 zTree 单选按钮/复选框功能 …… 232
　8.3.3 zTree 的拖拽功能 …… 235
　8.3.4 zTree 实现节点的增加、删除、修改功能 …… 237
8.4 JQGrid 表格控件 …… 241
　8.4.1 JQGrid 的原理 …… 241
　8.4.2 JQGrid 的安装 …… 242
　8.4.3 JQGrid 的参数 …… 242
　8.4.4 JQGrid 中 ColModel 的 API …… 246
　8.4.5 JQGrid 的代码格式 …… 248
8.5 JQGrid 实例 …… 249
习题 …… 259
综合应用 …… 259

第 9 章 AJAX 技术 …… 261
9.1 AJAX 概述 …… 261
9.2 原理简介 …… 262
　9.2.1 创建对象 …… 262
　9.2.2 发送请求 …… 262
　9.2.3 获取响应 …… 267
　9.2.4 onreadystatechange 事件 …… 269
9.3 JQuery AJAX …… 269
　9.3.1 load() …… 269
　9.3.2 get() …… 274
　9.3.3 post() …… 277
9.4 综合实例 …… 281
　9.4.1 多属性查询 …… 281
　9.4.2 分页显示 …… 286
习题 …… 292
综合应用 …… 292

第 10 章 服务器（IIS）的配置与使用 …… 293
10.1 IIS 简介 …… 293
10.2 IIS 安装 …… 293
10.3 IIS 的属性与配置 …… 296
10.4 工程在 IIS 上的发布 …… 298
习题 …… 301
综合应用 …… 302

参考文献 …… 303

第1章

MVC 4 简介

MVC 模式最早由 Trygve Reenskaug 在 1974 年提出，是施乐帕罗奥多研究中心（Xerox PARC）在 20 世纪 80 年代为程序语言 Smalltalk 发明的一种软件设计模式。本章主要介绍一般页面制作时，需要掌握的知识、思想、结构，并从 HTML、JavaScript、ASP.NET 的应用和 MVC 的原理出发，逐步深入讲解有关知识。

1.1 基础知识

1.1.1 HTML

HTML 作为一种描述网页的语言，是超文本标记语言（Hyper Text Markup Language）的简称。HTML 是由标签（Tag）和属性（Attribute）组成，浏览器读取 HTML 标签和属性，进而解析成网页的形式显示出来，因此我们能在浏览器中方便地查看网页。

HTML 和我们所说的 C 语言、Java 有很大的不同，它是一种标记语言。从浏览器中可以看到文字、图片、视频，还可以听到音频，这些都是网页的内容，只是内容的种类不一样，HTML 就是通过在文本文件中添加标记的方式，告诉浏览器该如何显示需要显示的内容，比如样式如何安排、图片的位置等。

下面通过一个例子了解一下 HTML。我们新建一个文本文档，命名为"测试"，并填入如下内容。

【例 1.1】 HTML 测试示例。

```
1    <html>
2        <head>
3            <title>
4                测试
5            </title>
6        </head>
7        <body>
8            这是一个简单的测试
9        </body>
10   </html>
```

然后将其扩展名改成"html",双击可以看到是通过浏览器打开的该文件,我们可以看到一个很简单的网页,如图1.1所示。

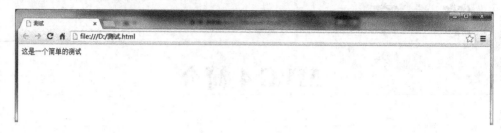

图1.1　HTML测试

HTML元素是由一对尖括号包围的关键词,一般成对出现,比如<hl>和</hl>。每个标签对中,第一个称为起始标签(开放标签),第二个称为结束标签(闭合标签)。某些没有内容的HTML元素称为空元素,即在起始标签中是关闭的,没有结束标签,用起始标签添加斜杠表示,如
。

目前,HTML不区分大小写,即<HTML>、<Html>、<html>是等效的,本章节中使用小写。大多数的HTML标签可以嵌套使用,同时HTML标签具有对应的属性和属性值。下面是一些常用的HTML标签及其对应的属性(w3school,2013)。

1. 基本标签

(1) <html></html>——HTML文档标签

表示文档类型为HTML类型,并且处于HTML文档的最外层,文档中的所有文本和HTML标签都包含在内。

(2) <head></head>——文档头部标签

描述文件的头部信息,包括文档的标题、与其他文档关系以及在Web中的位置等,故如定义页面中所有连接的基准URL的<base>标签、定义资源引用的<link>标签、用于定义元信息的<meta>标签以及文档标题的<title>标签等可以写在head部分,同样head标签里也可以包含JavaScript和CSS,其中JavaScript和CSS分别在第7章和1.1.2节中给出介绍。例如,在<meta>标签里加入关键词,就有助于搜索引擎搜索到该网页。

(3) <title></title>——文档标题标签

表示文档的标题,通常将显示在浏览器窗口的标题栏或者状态栏上,若将文档加入收藏夹时,文档标题也作为其默认名称。

(4) <body></body>——文档主体标签

表示文档的主体,包含文档的文本、超链接、图像、表格、列表等内容。

(5) <!--注解-->——注释标签

在HTML文档中,有时为了方便自己和用户了解文件内容,需要编写注释文件。注释文件的内容并不会显示在浏览器中。

2. 排版标签

(1) <p></p>——段落标签

表示一个段落,在浏览器中以段落的格式显示标签之间的内容。

<p></p>具有align属性,用以设置段落的对齐方式。

例如:<p align="left">你好</p>,表示"你好"在浏览器中以左对齐的方式显示。

(2)
——换行标签

表示在文档中插入换行符。

(3) <hr/>——水平线标签

表示在 HTML 文档中加入一条水平线。<hr>具有 align、size、color、width 和 noshade（无阴影）属性，分别用以设置水平线的水平对齐方式、水平线的高度（单位为像素）、颜色、宽度和立体效果。

例如：<hr align = "center" color = "red" width = "33" size = "12" noshade />，表示一条宽 33 像素、高 12 像素、无阴影并且居中对齐的红色水平线。

(4) <div></div>——块定义标签

表示文档中的分区或者块，作为块级元素标签，<div></div>标签具有 id、class、align、style 等属性，分别标识元素唯一的 ID、类名、水平对齐方式以及元素的行内样式。下面通过一个例子来展示如何设置 div 标签。

【例 1.2】div 标签设置示例。

```
1      < div style = " border: 5px solid # 000000; padding: 10px; margin: 15px;
2      background-color: #FFFF00; width: 100px; height: 50px; text-align: center;">
3          第一个 div
4      </div>
5      < div style = " background-color: #00FFFF; font-size: 28px; text-decora
6      tion: underline; font-family:
7      楷体; color: #FF0000;">
8          第二个 div
9      </div>
```

代码中设置的第一个边框为黑色，边框宽 5 像素（px），外边距 10 像素，内边距 15 像素，宽 100 像素，高 50 像素，背景色为黄色，居中显示"第一个 div"。border、padding 和 margin 的关系如图 1.2 所示。

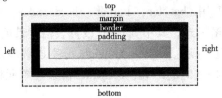

图 1.2　border、padding 和 margin 的关系

"第二个 div"为红色楷体，大小为 28 像素，带下画线，背景呈水蓝色。

页面运行结果如图 1.3 所示。

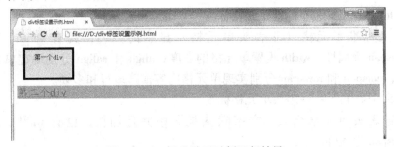

图 1.3　div 标签设置示例运行结果

3. 文本标签

(1) <h*n*> </h*n*>——标题标签

用于设置文本中标题的大小，*n* 可以取值 1~6，其中 <h1> </h1> 对应最大的标题，<h6> </h6> 对应最小的标题。标题标签间的文字显示为黑体，并自动插入一个空行。

(2) <font> </font>——文字格式标签

用于设置文本中文字的格式，具有 face、size、color 属性，分别表示文字的字体、大小、颜色。

例如：<font face = "宋体" size = "2" color = "red">文本文字</font>，表示显示为红色宋体且大小为 2 像素的文本文字。

(3) 特殊文字样式标签

HTML 中有设置文字显示为特殊格式的标签，表 1.1 列出一些常用的特殊文字样式标签。

表 1.1 常用的特殊文字样式标签和显示效果

标签名称	标签示例	显示效果
粗体	<b>你好</b>	**你好**
斜体	<i>你好</i>	*你好*
下画线	<u>你好</u>	你好
删除线	<s>你好</s>	你好
强调	<strong>你好</strong>	**你好**
上标	<sup>你好</sup>	你好
下标	<sub>你好</sub>	你好

4. 表格标签

(1) <table> </table>——表格标签

用于创建表格，使用 bgcolor、border、bordercolor、width、height、cellspacing、cellpadding 等属性来设置表格格式，其中 bgcolor 设置表格背景色；border 设置边框宽度，默认值为 0；bordercolor 设置边框颜色；width 与 height 分别设置表格的宽度和高度；cellspacing 设置单元格的间距，即单元格之间的距离；cellpadding 设置单元格的边距，即字体与单元格边框的距离。

(2) <tr> </tr>——单元格行标签

用于定义表格的行，具有 align、valign 等属性，分别设置行的水平对齐方式和垂直对齐方式等。

(3) <td> </td>——标准单元格标签

用于定义标准单元格（非表头单元格）显示的内容，具有 width、align、valign、colspan、rowspan 等属性。width 设置单元格的宽度；align 和 valign 分别规定单元格水平和垂直对齐方式；colspan 和 rowspan 分别实现单元格内容横跨多行和多列。

(4) <th> </th>——表头单元格标签

用于设置表头单元格格式，文字默认是居中并且加粗，也有 width、align、valign、colspan、rowspan 等属性。

5. 表单标签

（1）＜form＞＜/form＞——表单标签

用以创建所有表单，＜form＞具有 action、method 和 target 属性。action 是必选属性，定义表单提交的位置（相对地址或绝对地址）；method 定义了表单数据传送到服务器的处理方法，即 get 及 post，在以 get 方式提交时，客户端浏览器将每个表单域的 name 和 value 进行 URL encoding 处理，而以 post 方式提交表单时，客户端浏览器不会自动处理 URL 中的非法字符（http：//www.nowamagic.net/librarys/veda/detail/182，2013）；target 属性用来指定目标窗口。

举个简单的例子：＜form method = "post" action = "http：//www.ustb.edu.cn/"＞＜/form＞，则表示将表单处理结果通过 post 的方式传送到 http：//www.ustb.edu.cn 上。

（2）＜input type = "type" name = "name"＞——输入区标签

用来定义一个用户输入区，不同的 type 属性值决定不同类型的输入区域，具体内容见表 1.2。

表 1.2　type 取值和相应的输入区域类型

type 取值	输入区域类型
＜input type = "text" name = "NAME" value = "value" size = "size" maxlength = "length"＞	文本输入区域，value 表示初始值，size 表示宽度，length 表示最大输入字符数
＜input type = "submit" name = "NAME"＞	提交表单内容（到服务器）的按钮
＜input type = "reset" name = "NAME"＞	重置表单内容的按钮
＜input type = "checkbox" name = "NAME" value = "value" checked＞	复选框，checked 表示默认选中取值为 value 的选项
＜input type = "radio" name = "NAME" value = "value" checked＞	单选框，checked 表示默认选中取值为 value 的选项
＜input type = "password" name = "NAME"＞	密码输入框，默认显示为"****"
＜input type = "image" name = "NAME" src = "name.gif"＞	将图像引入页面，用于替换 submit 按钮，但是需要添加 onclick 事件

（3）＜select＞＜/select＞——列表框标签

用以创建一个列表框。＜select＞＜/select＞标签具有 multiple、name、size 等属性，其中 multiple 属性表示可以有多个选择，name、size 属性分别用来设置列表框名字和高度。

＜select＞＜/select＞标签中的＜option＞标签代表一个选择项。＜option＞具有 selected 和 value 属性，＜option selected＞表示默认的选项，value 表示该选项的值。

【例 1.3】一个基本的列表框的代码。

```
1    <select name = "水果"  size = "6">
2        <option>苹果</option>
3        <option selected>香蕉</option>
4        <option>梨</option> </select>
```

创建的列表框默认选择香蕉，效果如图 1.4 所示。

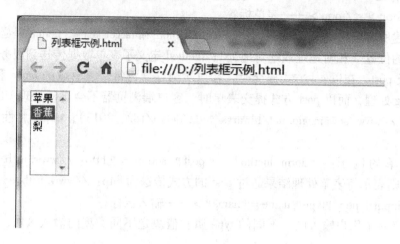

图 1.4 列表框示例

（4）＜textarea＞＜/textarea＞——文本区域标签

用以创建一个文本区域，＜textarea＞＜/textarea＞标签具有 name、cols、rows 属性，分别表示文本区域的名称、列数和行数。

6. 列表标签

（1）＜ul＞＜/ul＞——无序列表标签

用以创建一个无序的项目列表。＜ul＞＜/ul＞标签具有 type 属性，可以设置项目符号的样式。

（2）＜ol＞＜/ol＞——有序列表标签

用以创建一个有序的编号列表。＜ol＞＜/ol＞标签具有 type 和 start 属性，type 属性可以改变编号样式本身；start 属性则允许改变开始值（默认从"1"开始对有序列表的条目进行编号）。

（3）＜li＞＜/li＞——列表的项目标签

用以创建一个列表项。＜li＞＜/li＞标签必须处在＜ol＞＜/ol＞标签或＜ul＞＜/ul＞标签之间。

7. 其他标签

（1）＜img＞——图像标签

用以向网页中插入一幅图像，通过 src 属性指定的路径（绝对路径或相对路径）从网页上链接到图像。此外，＜img＞标签还具有 alt、align、border、width、height 等属性，分别用以设置图像的描述性文本（即当浏览器无法显示图像或者鼠标移动到图像上时显示的内容）、对齐方式、周围边框，以及图像的宽度和高度等。

例如：＜img src = "/i/tulip.jpg"alt = "郁金香" width = "120" height = "100"/＞，表示插入一张格式为 jpg，名称为 tulip，宽为 120 像素，高为 100 像素，文字信息为"郁金香"的图片。

（2）＜a＞——链接标签

用以创建超链接。＜a＞标签具有 href、name、id 等属性，其中 href 属性指向另外一个文档的链接（或超链接）；name 或 id 属性实现了页面的跳转。

例如：< a href = "http：//www.ustb.edu.cn" >北京科技大学首页</a >，表示创建一个链接到北京科技大学首页的超链接。

上述内容简单介绍了一些常用的 HTML 标签，其他一些重要的标签的语法可以参考 w3school 网站中 HTML 的相关内容。

1.1.2 CSS

CSS（Cascading Style Sheets）即"层叠样式表"或"级联样式表"，也简称为"样式表"。CSS 的引入分离了网页内容与视觉，因此页面维护工作、网页搜索都变得更加容易，而且增强了页面在不同媒介之间的呈现效果。

1. CSS 语法

CSS 定义的基本格式为：

选择符 {属性：属性值}

其中选择符可以是多种形式的，同定义样式的 HTML 标记，如 body、table 等；对于多个属性的情况，需要用分号将各属性分隔开；若属性值是由多个单词组成的，则需要给该值加引号。

【例1.4】一个简单的 CSS 示例。

```
1      P
2      {
3          Color: red;
4          Font-family: "Times New Roman"
5      }
```

上面这段代码的效果为：页面中段落间的文字为红色 Times New Roman 字体。

2. CSS 常用属性及属性值

CSS 的字体、颜色、文本属性及对应的属性值见表1.3。

表1.3 CSS 常用属性及属性值

	属性	属性描述	属性值
字体	Font-family	字体	所有可用字体
	Font-style	是否斜体	normal、italic、oblique
	Font-variant	是否小写	normal、small-caps
	Font-weight	字体粗细	normal、bold、bolder、lither
	Font-size	字体大小	absolute-size、relative-size、length 等
颜色	Color	前景色	颜色值
	Background-color	背景色	颜色值
	Background-image	背景图片	图片路径
	Background-repeat	背景图案重复样式	repeat-x、repeat-y、no repeat
	Background-attachment	设置滚动	scroll、fixed
	Background-position	设置背景图案初始位置	percentage、length、top、left、right、bottom

(续)

属性		属性描述	属性值
文本	Word-spacing	定义单词间距	normal＜length＞（以长度为单位）
	Letter-spacing	定义字母间距	normal＜length＞（以长度为单位）
	Text-decoration	定义文字的"装饰"样式	none、underline、overline、line-through、blink
	Vertical-align	定义元素在垂直方向的位置	baseline｜super｜top｜text-top｜middle｜bottom｜text-bottom＜percentage＞
	Text-transform	使文本转化为其他形式	capitalize、uppercase、lowercase、none
	Text-align	定义文本对齐方式	left、right、center、justify
	Text-indent	定义文本首行缩进方式	＜length＞、＜percentage＞
	Line-height	定义文本行高	normal、＜number＞、＜length＞、＜percentage＞

【例1.5】CSS文本属性示例。

```
1    <p style="letter-spacing: 2cm; text-align: justify; text-indent: 4cm; line-
2    height: 17pt; color: red">
3        假如生活欺骗了你<br/>
4        不要悲伤 不要心急<br/>
5        忧郁的日子里需要镇静<br/>
6        相信吧 快乐的日子将会来临<br/>
7        心儿永远向往着未来<br/>
8        现在却常是忧郁<br/>
9        一切都是瞬息<br/>
10       一切都将会过去<br/>
11       而那过去了的<br/>
12       就会成为亲切的回忆<br/>
13   <p/>
```

上述代码片段中，字间距为2cm，文本对齐方式为两端对齐，缩进4cm，行高为17pt。运行效果如图1.5所示。

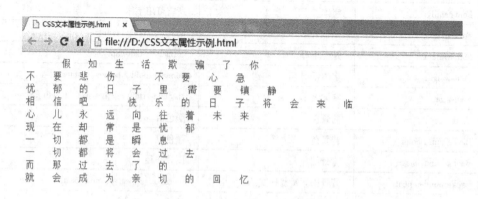

图1.5 CSS文本属性示例

3. CSS 的使用方法

CSS 语句是内嵌在 HTML 文档中的，故编写 CSS 的方法与 HTML 文档相似。因此，常用的 HTML 文本编辑工具，甚至记事本都可以用来编辑 CSS 文档，再按照下述提供的方法就能将独立编辑的 CSS 文档加入到 HTML 中。

方法一：在 < head > </head > 之间插入 < style type = "text/css" > … </style >，< style > 标签中的"type = "text/css""表示里面的代码是 CSS 样式。

方法二：采用 < style = "" > 的格式，将 CSS 样式写在 HTML 的行内。

方法三：将编辑好的 CSS 文档保存成 ".CSS" 文件，在 < head > </head > 中用如下的格式引用：

< head > < link rel = stylesheet href = "style.css" > </head >

在此使用了一个 < link > 标签，"rel = stylesheet" 表示链接的元素是一个样式表，"href = "style.css"" 标明链接文件的地址。

1.2 基于 WebForm 的 ASP.NET

1.2.1 .NET

.NET 作为 Microsoft XML Web services 的平台，因此，无论是基于何种类型的操作系统、设备和编程语言，XML Web services 都允许应用程序通过互联网（Internet）进行数据共享与通信。

.NET 不仅可以管理代码的执行，而且能为代码提供服务，扮演双重角色。

其中 .NET 提供的服务包括：①.NET 框架，是 .NET 提供的一种新的运行环境；②ASP.NET，是为创建 HTML 页面而提供的一种新的编程模型；③Windows 窗体，是为编写各种程序提供的新方法；④XML Web，为 Internet 服务器提供新的方法；⑤ADO.NET，支持数据库访问。

.NET 框架支持多语言开发，便于建立 Web 应用程序及 Web 服务，加强 Internet 上各应用程序之间通过 Web 服务的沟通。

.NET 框架主要包括三个主要组成部分：公共语言运行时（Common Language Runtime，CLR）、服务框架（Services Framework）及应用模板。

其中，CLR 管理了 .NET 中的代码，管理内存与线程；服务框架为开发人员提供了一套基于标准语言库的基类库，除了基类之外，包括接口、值类型、枚举和方法，可完成许多不同的任务，以简化编程工作；应用模板包括传统的 Windows 应用程序模板（Windows Forms）和基于 ASP.NET 的面向 Web 的网络应用程序模板（Web Forms 和 Web Services）。图 1.6 展示了各部件之间的关系。

1.2.2 ASP.NET

1. ASP.NET 简介

ASP（Active Server Page）全称为动态服务页面，结合了脚本、超文本和强大的数据库访问功能，并提供了众多的服务器端组件便于程序直接调用。

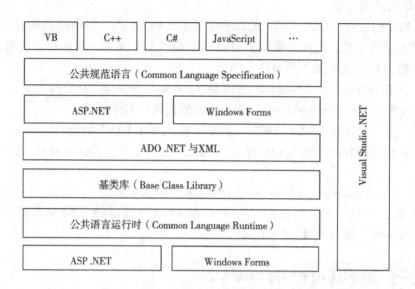

图1.6 Visual Studio .NET 框架部件关系

ASP.NET 是微软公司推出的用于 Web 开发的全新框架,底层采用.NET 框架,使用 HTML、CSS、JavaScript 以及服务器脚本来构建网页和网站。ASP.NET 网页在浏览器端向用户提供信息,然后通过服务器端代码实现应用程序的逻辑。

2. ASP.NET 对象

ASP.NET 的基本对象包括 Request、Response、Application、Session、Cookie 等,是实现用户交互功能的基础。Request 对象用于获取用户输入的信息,Response 对象用于浏览器的输出。Application 对象用于存储一个特定应用程序中所有用户的信息,使得同一个应用程序中的所有页面均有效,而 Session 对象用于存储诸如姓名、ID 或者参数等单一用户的数据,对同一个应用程序中的所有页面有效。Application 与 Session 对象均可供多个 ASP 文件调用。Cookie 使用范围较广,常用的属性和方法见表1.4。

表1.4 Cookie 对象的属性和方法

	名称	说 明
属性	Name	获取或设置 Cookie 的名称
	Value	获取或设置 Cookie 的值
	Expires	获取或设置 Cookie 的过期日期和时间
	Version	获取或设置此 Cookie 符合的 HTTP 状态维护版本
方法	Add	新增一个 Cookie 对象
	Clear	清除 Cookie 集合内的变量
	Get	通过变量名或索引得到 Cookie 的变量值
	GetKey	以索引值来获取 Cookie 的变量名称
	Remove	通过 Cookie 变量名删除 Cookie 变量

1.2.3 WebForm 的创建方法

在 Visual Studio 上创建一个 ASP.NET 网站,下面是创建过程的详细步骤。

Step01:启动 Visual Studio。

Step02:单击"文件",选择"新建",然后在弹出的快捷菜单中选择"网站",单击进入,如图 1.7 所示。

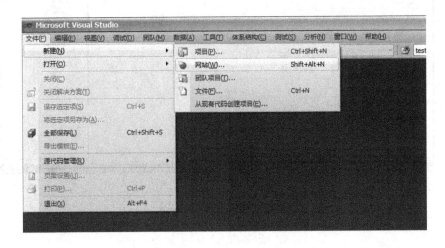

图 1.7 新建网站

Step03:在弹出的"新建网站"对话框中,选择"Visual C#",然后选择"ASP.NET 网站"。网站名称设置为"Test_WebSite1",存储位置设置为"D:\Test_WebSite1"(可自行设定),如图 1.8 所示。

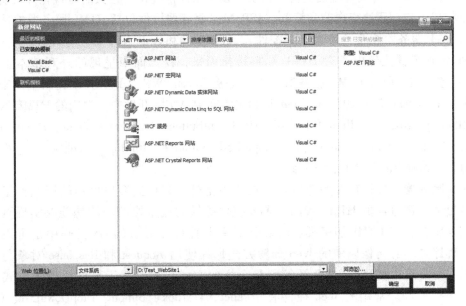

图 1.8 创建一个 ASP.NET 网站

Step04：单击"确定"按钮，这样我们就创建了一个WebForm，如图1.9所示。

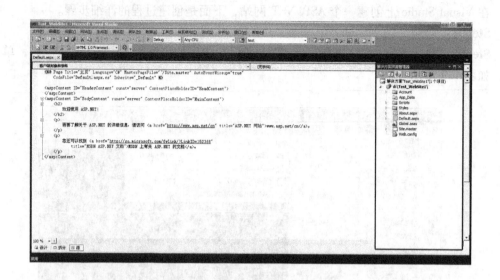

图1.9 创建后的WebForm

1.2.4 ASP.NET 服务器控件

　　ASP.NET程序架构是建立在通用语言环境上的，可以利用ASP.NET建立基于Web的B/S（浏览器/服务器）结构应用程序，并且这种架构还有微软公司的Visual Studio 2008（及以上版本）集成开发环境的支持，由于可以使用控件提供的强大的可视化开发功能，因此，开发Web应用程序变得非常简单。ASP.NET服务器控件包括HTML服务器控件（基础控件）、Web服务器控件（简称Web控件）、数据控件和ASP.NET移动控件。

　　HTML服务器控件由普通HTML标签转换而来，在页面中呈现的外观基本上与普通HTML标签一致。所有HTML服务器控件都是从System.Web.UI.Control类派生而来的，并且都包含在System.Web.UI.HtmlControls命名空间中。实际开发中，常用的HTML服务器控件有HtmlInputText、HtmlInputPassword、HtmlInputButton、HtmlInputSubmit、HtmlInputReset、HtmlInputImage、HtmlInputhidden、HtmlInputRadioButton和HtmlInputCheckBox。这些控件的基本功能与其对应的HTML标签一致。

　　Web服务器控件运行在服务器端，在初始化时，会根据客户使用的浏览器工具的不同而给出适合浏览器的HTML代码。Web控件提供的最重要的功能是能够实现页面与后台的交互服务，与HTML服务器控件相比，Web控件在后台的CS代码中是可以直接访问的，程序员不必再像以前的Web开发方式那样使用Request和Response对象与页面进行交互，而是可以使用和RAD开发工具一样的方式，通过基于事件模型的方式使用页面上的各种控件。常用的Web控件有：Label、TextBox、Button、DropDownList、Calendar等。

　　数据控件包括数据访问控件和数据绑定控件。数据访问控件又称数据源控件，主要用于访问数据库，简化数据库访问操作。而数据绑定控件可以从本质上改变代码的质量，为.NET框架提供更高级别的可重用性，其控件主要有：GridView、DetailsView、FormView、DataList、ListView、DataPager 等。

下面是一个应用数据控件对表进行增加、删除、修改和查询的例子。

【**例1.6**】窗体上有两个文本框（txtNo、txtName）和4个按钮（分别用以增加、删除、修改和查询）。

```
1    using System;
2    using System.Collections.Generic;
3    using System.ComponentModel;
4    using System.Data;
5    using System.Drawing;
6    using System.Text;
7    using System.Windows.Forms;
8    using System.Data.SqlClient;
9    namespace test
10   {
11       public partial class Form2 : Form
12       {
13           public Form2()
14           {
15               InitializeComponent();
16           }
17           DataSet ds;
18           SqlDataAdapter ada;
19           string sql = null;
20   SqlCommand cmd;
21   //查询
22           private void btnSearch_Click(object sender, EventArgs e)
23           {
24               ShowView();
25           }
26           private void ShowView()
27           {
28               ds = new DataSet("myschool");
29               if (ds.Tables.Count > 0)
30               ds.Tables.Clear();
31               sql = "SELECT * FROM classinfo";
32               ada = new SqlDataAdapter(sql, DBConn.conn);
33               ada.Fill(ds);
34               this.dataGridView1.DataSource = ds.Tables[0];
35           }
```

```csharp
36          //添加
37                  private void btnAdd_Click(object sender, EventArgs e)
38                  {
39  sql = string.Format("INSERT classInfo VALUES ({0},'{1}')", int.Parse(txt
40  No.Text), txtName.Text);
41                      cmd = new SqlCommand(sql, DBConn.conn);
42                      DBConn.conn.Open();
43                      int count = cmd.ExecuteNonQuery();
44                      if (count > 0)
45                      {
46                          MessageBox.Show("添加成功!");
47                      }
48                      else
49                      {
50                          MessageBox.Show("添加失败!");
51                      }
52                      DBConn.conn.Close();
53                      ShowView();
54                  }
55          //删除
56                  private void btnDelete_Click(object sender, EventArgs e)
57                  {
58  sql = string.Format("delete classinfo WHERE classno = {0}", int.Parse(txtNo.Text));
59                      cmd = new SqlCommand(sql, DBConn.conn);
60                      DBConn.conn.Open();
61                      int count = cmd.ExecuteNonQuery();
62                      if (count > 0)
63                      {
64                          MessageBox.Show("删除成功!");
65                      }
66                      else
67                      {
68                          MessageBox.Show("删除失败!");
69                      }
70                      DBConn.conn.Close();
71                      ShowView();
72                  }
73          //修改
74                  private void btnUpdate_Click(object sender, EventArgs e)
75                  {
76                      SqlCommandBuilder bl = new SqlCommandBuilder(ada);
77                      ada.Update(ds);
78                      ShowView();
```

```
79                }
80            }
81        }
82    //按钮的触发函数
83        using System;
84        using System.Collections.Generic;
85        using System.ComponentModel;
86        using System.Data;
87        using System.Drawing;
88        using System.Text;
89        using System.Windows.Forms;
90        using System.Data.SqlClient;
91        namespace nycx
92        {
93            public partial class Form1 : Form
94            {
95                public Form1()
96                {
97                    InitializeComponent();
98                }
99                DataSet ds;
100               SqlDataAdapter ada;
101               string sql = null;
102               SqlCommand cmd;
103               private void Form1_Load(object sender, EventArgs e)
104               {
105                  ShowView();
106               }
107               private void button1_Click(object sender, EventArgs e)
108               {
109                  ShowView();
110               }
111               private void ShowView()
112               {
113                   SqlConnection con = DBconn.conn;
114                   ds = new DataSet("myschool");
115                   if (ds.Tables.Count > 0)
116                       ds.Tables.Clear();
117    sql = "SELECT UserID, UserName, Password, TrueName, DeptCode FROM dbo.[User]";
118                   ada = new SqlDataAdapter(sql, con);
119                   ada.Fill(ds);
120                   this.dataGridView1.DataSource = ds.Tables[0];
121               }
```

```
122            private void ss(String dd)
123            {
124                SqlConnection con = DBconn.conn;
125                ds = new DataSet("myschool");
126                if (ds.Tables.Count > 0)
127                    ds.Tables.Clear();
128                sql = string.Format ( "SELECT UserID, UserName, Password, True
129        Name, DeptCode FROMdbo.[User] where UserName like '%{0}%'",dd);
130            ada = new SqlDataAdapter(sql, con);
131                ada.Fill(ds);
132                this.dataGridView1.DataSource = ds.Tables[0];
133            }
134            private void button2 _ Click(object sender, EventArgs e)
135            {
136                ss(text1.Text);
137            }
138        }
139    }
```

ASP.NET 通过验证控件可以轻松地实现对用户输入的验证，而且还可以选择验证在服务器端进行还是在客户端进行，从而让程序员可以把主要精力放在主程序的设计上。验证控件包括必填字段验证控件 RequiredFieldValidator、范围验证控件 RangeValidator、正则表达式验证控件 RegularExpressionValidator 和自定义验证控件 CustomValidator。

1.3 MVC 概念与原理

1.3.1 MVC 的概念

MVC（Model-View-Controller，即模型-视图-控制器模式）是软件工程中的一种软件架构模式，也是 ASP.NET 的开发模式之一。它把软件系统分为三个基本部分：Model（模型）、View（视图）和 Controller（控制器），每一部分都相对独立，职责单一。

1.3.2 MVC 的工作原理

1. 核心部件

（1）Model

模型，作为应用程序的主体部分，封装了与应用程序的业务逻辑相关的数据以及对数据的处理方法，包括数据格式验证以及数据库的操作等。

（2）View

视图，即为与用户交互的界面，不仅可以接收用户的输入数据，也可以向用户展示相关的数据。视图一般不涉及程序上的逻辑，使得页面独立于逻辑。

（3）Controller

控制器主要是进行逻辑处理，控制实体数据在视图上的展示，并调用模型处理业务请求。总之，控制器能在不同的层之间控制应用程序的流程，起到了组织的作用。

2. 运行机制

在 MVC 模式中，Web 用户向服务器提交的所有请求都由控制器接管。接收到请求之后，控制器负责决定应该调用哪个模型来进行处理；然后模型根据用户请求进行相应的业务逻辑处理，并返回数据；最后控制器调用相应的视图来格式化模型返回的数据，并通过视图呈现给用户，如图 1.10 所示。

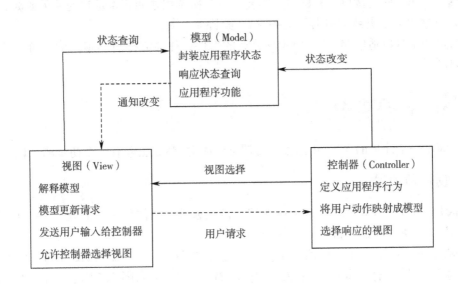

图 1.10　MVC 运行机制

从其运行机制中不难发现这三个部件之间既相对独立，又相互联系。

1.3.3　MVC 架构的优缺点

我们已经对 MVC 架构有了一些基本的了解，下面开始深入探讨 MVC 架构的优缺点。

1. MVC 架构的优点

（1）高内聚，低耦合

各模块之间依赖性小，层层分离，在这些元素之间提供松散耦合。而各个模块内部的数据代码依赖性又是高度聚合的。

（2）提高代码重用率

MVC 的低耦合性决定了它的高度重用性。同一个模型也可以被多个视图重用。

（3）提高开发效率，加快了程序开发，有利于团队开发

MVC 应用程序的这三个主要组件之间的松散耦合也可促进并行开发，这样可以物尽其用，分工合作。

（4）提高程序的可维护性

因为模型与控制器和视图相分离，所以修改数据层和业务规则就变得比较容易。

2. MVC 架构的缺点

但是 MVC 架构也不是绝对完美的，存在以下的缺点。

（1）增加了系统结构和实现的复杂性

对于简单的界面，严格遵循 MVC，将模型、视图与控制器分离，反而会增加结构的复杂性。

（2）视图与控制器间的连接过于紧密

视图与控制器是相互分离，但又紧密联系的部件，因此会妨碍独立重用。

（3）视图对模型数据的访问效率较低

对模型操作不同的接口，视图可能需要经过多次调用才能获得足够的显示数据。

（4）某些界面工具或构造器不支持 MVC 架构

改造这些工具以适应 MVC 需要建立分离的部件，往往需要付出很高的代价，从而造成使用 MVC 的困难。

1.4 初识 MVC 4

下面演示一个简单的 MVC 4 工程，希望读者能对 MVC 工程有一个初步的认识。

1.4.1 创建新项目

Step01：打开 Visual Studio Express 2012，依次选择"文件"→"新建"→"项目"，单击"项目"，出现"新建项目"对话框。

Step02：在"新建项目"对话框中，依次选择"Web"→"ASP.NET MVC 4 Web 应用程序"，填写项目名称，此处命名为"MyFirstMVC 4"（见图 1.11），单击"确定"按钮。

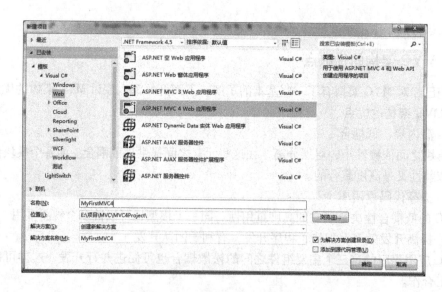

图 1.11 "新建项目"对话框

Step03：在"新的 ASP. NET MVC 4 项目"对话框中，选择"Internet 应用程序"，使用 Razor 作为默认视图引擎（见图 1.12）。

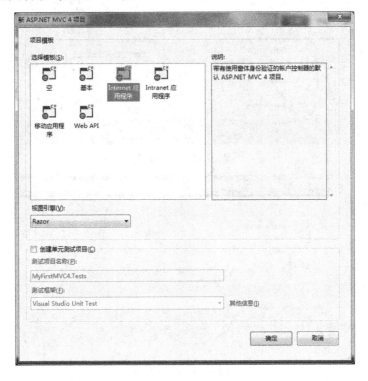

图 1.12 "新 ASP. NET MVC 4 项目"对话框

Step04：单击"确定"按钮后，使用默认的模板就成功新建了一个 MVC 4 项目。接下来，依次选择"调试"→"启动调试"，或者直接按 < F5 > 键启动项目，则能得到如图 1.13 所示的效果。

图 1.13 调试后生成的页面

本例中的端口号为 3032，读者在实际操作中端口号可能会有所不同。细心的读者可能就会发现在默认的程序中，已经实现了登录（登录界面见图 1.14）、注册（注册界面见图 1.15）等功能。

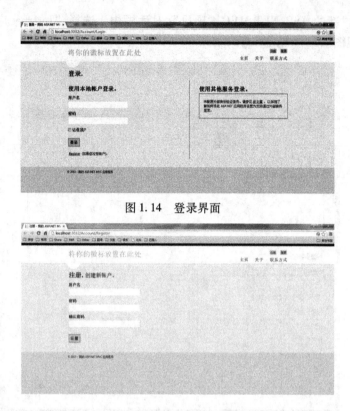

图 1.14 登录界面

图 1.15 注册界面

Step05：在项目创建完成后，我们不妨看一下"解决方案资源管理器"（见图 1.16）。

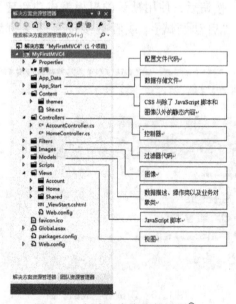

图 1.16 解决方案资源管理器[⊖]

⊖ http：//www.cnblogs.com/mzwhj/archive/2013/01/30/2883248.html

Content 放置网站的静态内容，如 CSS、image（图像）、Flash 等；Controllers 用于存放创建的控制器；Models 用于存放 Model 层的内容；Views 即为视图层，存放视图；Global.asax 定义网址路由。

1.4.2 MVC 4 项目

下面以 MyStudent 数据库为例进行讲解。该数据库包含 Student（Sno，Sname，Ssex，Sage，Sdept）、Course（Cno，Cname，Cpno，Ccredit）和 SC（Sno，Cno，Grade）三张表，创建一个新的 MVC 4 项目。

1. 创建数据模型

Step01：在"解决方案资源管理器"中选择 Models 文件夹，右击它，在弹出的快捷菜单中依次选择"添加"→"新建项"，如图 1.17 所示，则出现"添加新项-MVCStudent"对话框，如图 1.18 所示。

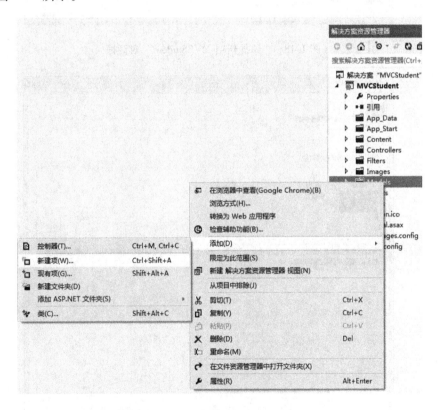

图 1.17 新建项步骤引导

Step02：在"添加新项-MVCStudent"的对话框中，依次选择"数据"→"ADO.NET 实体数据模型"，并将其改名为"MyStudent.edmx"，如图 1.18 所示，然后单击"添加"按钮。

Step03：在出现的"实体数据模型向导"对话框中，选择"从数据库中生成"，如图 1.19 所示，然后单击"下一步"按钮。

Step04：单击"新建连接"（若不是第一次连接这个数据库，在新建连接的下拉列表框

图 1.18 "添加新项-MVCStudent"对话框

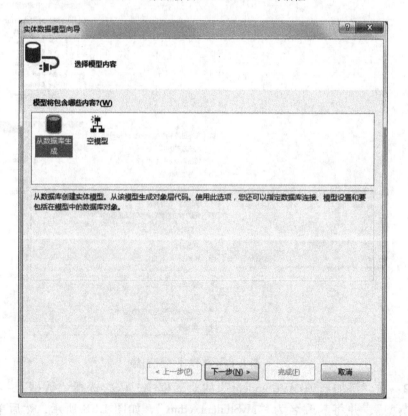

图 1.19 选择从数据库生成

中可以找到所要的连接,则不必新建连接),在出现的"连接属性"对话框(见图 1.20)中,填入用户的数据库名,也可在"选择或输入数据库名称"的下拉列表中选择用户要连接的数据库。

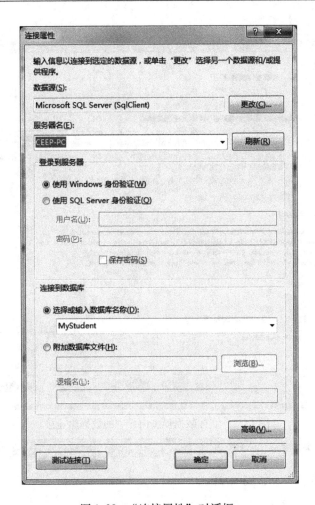

图 1.20 "连接属性"对话框

单击"测试连接"按钮,若显示"测试连接成功"(见图 1.21),则连接可用。

图 1.21 测试成功

单击"确定"按钮,可看到"实体数据模型向导"对话框中出现刚才连接的数据库,并且自动命名为 MyStudentEntities(见图 1.22),然后单击"下一步"按钮。

注意:在创建实体数据类型时,实体数据类型是有单复数之分的,数据库和表的类型使用复数,实例使用单数,因此这里是有 s 的。

Step05:在如图 1.23 所示的对话框中,分别选中"表"和"确定所生成对象名称的单复数形式"复选框,然后单击"完成"按钮。

图 1.22　实体数据模型向导中的数据库连接

图 1.23　实体数据模型向导中的选择数据库对象

需要注意的是，如果数据库中有视图或者存储过程和函数时，也需要选中对应的复选框。本例中的数据库中仅有表。

单击 MyStudent.edmx 文件，则会显示如图 1.24 所示的关系图。因为本例中没有建立外码，所以图 1.24 中的三个表没有联系。

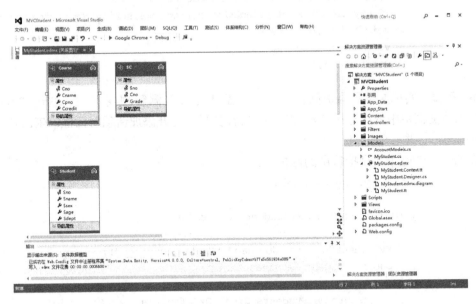

图 1.24　MyStudent.edmx 文件

Step06：实现接口层对数据的封装，此处以 Student 表为例。

Step06-1：写一个抽象的数据访问接口。在"解决方案资源管理器"中，右击"Models"，在弹出的快捷菜单中依次选择"添加"→"新建项"，打开"添加新项-MyStudent"对话框，依次选择"代码"→"接口"，并为接口命名为 IStu.cs（见图 1.25），然后单击"添加"按钮。

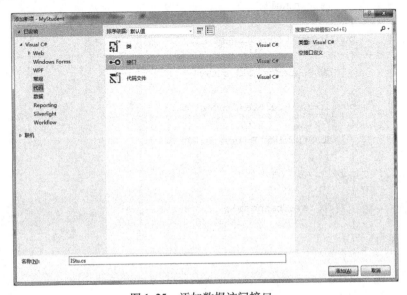

图 1.25　添加数据访问接口

在 IStu.cs 中输入下面这段代码。

```
1    interface IStu
2    {
3        Student GetStudent(string id);
4        IQueryable<Student> FindAllStudent();
5        IQueryable<Student> FindByKey(string name, string cla, string sno);
6        void Add(Student stu);
7        void Delete(Student stu);
8        void Save();
9    }
```

Step06-2：实现这个数据访问接口。在"解决方案资源管理器"中，右击"Models"，在弹出的快捷菜单中依次选择"添加"→"新建项"，打开添加新项对话框，依次选择"常规"→"类"，并取名为 Stu.cs。

在 Stu.cs 中逐个实现接口层的函数。

```
1    namespace MyStudent.Models
2    {
3        public class Stu : MyStudent.Models.IStu
4        {
5            MyStudentEntities db = new MyStudentEntities();
6            public Student GetStudent(string id)
7            {
8                return db.Students.SingleOrDefault(s => s.Sno == id);
9            }
10           public IQueryable<Student> FindAllStudent()
11           {
12               return db.Students;
13           }
14           public IQueryable<Student> FindByKey(string name, string cla, string sno)
15           {
16               return null;
17           }
18           public void Add(Student stu)
19           {
20               db.Students.Add(stu);
21           }
22           public void Delete(Student stu)
23           {
24               db.Students.Remove(stu);
25           }
26           public void Save()
27           {
28               db.SaveChanges();
29           }
30       }
31   }
```

2. 添加控制器及视图

Step01：下面开始新建 MyStudentController 控制器。首先用鼠标右键单击 Controllers 文件夹，在弹出的快捷菜单中依次选择"添加"→"控制器"，如图 1.26 所示。

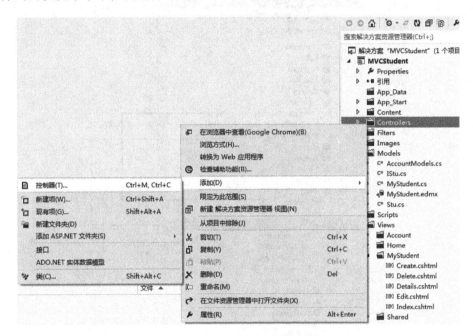

图 1.26　添加控制器步骤引导

在弹出的"添加控制器"对话框中进行设置，如图 1.27 所示。

图 1.27　"添加控制器"对话框

单击"添加"按钮之后，在 Views \ MyStudent 文件夹下会自动创建 Create.cshtml、Delete.cshtml、Details.cshtml、Edit.cshtml 和 Index.cshtml 文件，如图 1.28 所示。不难发现，控制器中的文件夹总能在视图中找到与之相对应的文件夹。

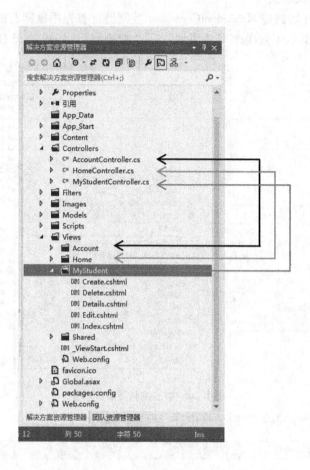

图 1.28 MyStudent 文件夹

Step02：编写对应视图的代码。

修改 Index.cshtml 中最后一对 <td> </td> 标签之间的代码，如下：

```
1    <td>
2        @Html.ActionLink("Edit", "Edit", new { id=item.Sno}) |
3        @Html.ActionLink("Details", "Details", new {id=item.Sno }) |
4        @Html.ActionLink("Delete", "Delete", new {id=item.Sno })
5    </td>
```

修改 Details.cshtml 中最后一对 <p> </p> 标签之间的代码，如下：

```
1    <p>
2        @Html.ActionLink("Edit", "Edit", new { id=Model.Sno }) |
3        @Html.ActionLink("Back to List", "Index")
4    </p>
```

单击 VS 中的运行按钮运行网页程序，在地址栏添加"/MyStudent"，即可实现增加、删除、修改以及查看详细信息等功能。

图 1.29 展示了"Index"页面，选择第一条记录的"Details"查询李名的详细信息，即

进入"Details"页面，如图 1.30 所示。

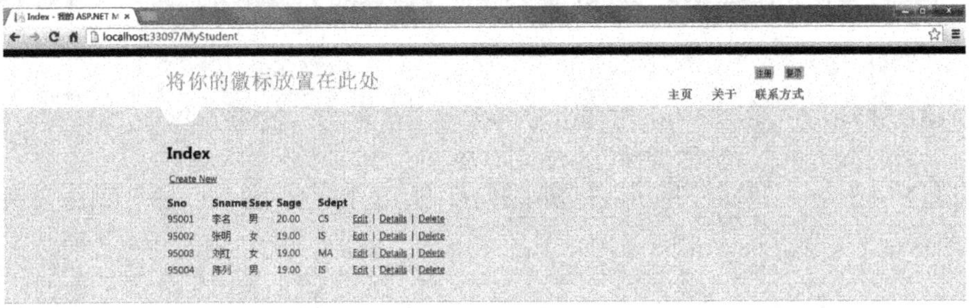

图 1.29　"Index"页面

图 1.30　"Details"页面

选择"Edit"，则进入"Edit"页面，如图 1.31 所示，将其年龄修改为 21，单击"Save"按钮后，自动跳转到更新后的"Index"页面，如图 1.32 所示，此时李名的年龄由 20 变成了 21。

图 1.31　"Edit"页面

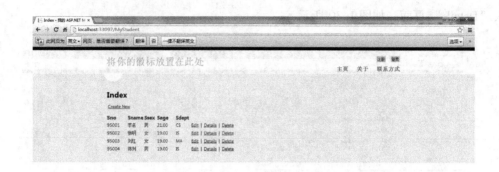

图 1.32 编辑信息之后的"Index"页面

单击页面左上方的"Create New",转入"Create"页面,可添加相应的记录,如图 1.33 所示。单击"Create"按钮后,自动跳转到更新后的"Index"页面,如图 1.34 所示。

图 1.33 "Create"页面

图 1.34 增加记录后的"Index"页面

选中最后一条记录,单击"Delete",在弹出的对话框中单击"确定"按钮,即可成功删除该条记录。

关于数据模型、控制器及视图的具体内容将在以后的章节中详细介绍。

※习　题

1. MVC 的本质是什么？为什么要提出 MVC 框架？
2. 编写 CSS，要求：修改 1.4.2 节图 1.29 中显示的表格为虚线红色外框；表头加粗，蓝色，3 号宋体，居中；内容黑色，5 号宋体（数字或字母为 Time New Roman），中文和英文居中显示，数字居右显示。
1) 使用 \<head\> \</head\> 中的 \<style\> \</style\> 格式编写。
2) 使用在组件内部的 Style 编写。
3) 使用 CSS 文件（*.CSS）进行编写。
3. 完成"Create"页面中 Create 的真正添加功能，并在显示页面显示出来（显示结构按照数据库中数据插入的先后顺序显示）。
4. 修改显示页面显示数据的条数，要求：分别显示前 10、20 条数据和后 10、20 条数据（时间上，早插入的数据排在后面）。
1) 在后台直接返回 10、20 条排序号的数据。
2) 后台返回所有数据，在前台进行数据的筛选显示。

※综合应用

基于默认建立的 MVC 4 基本工程，按照以下内容进行修改。

工程共分为 Home、About 和 Contact 三部分，其中 Home 页面如图 1.35 所示。

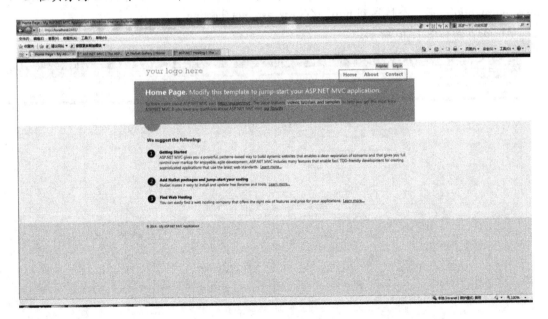

图 1.35　Home 页面内容

其中，Home 页面内容为：

Home Page. Modify this template to jump-start your ASP. NET MVC application.

这段文字为白色标题。

To learn more about ASP. NET MVC visit http：//asp. net/mvc. The page features videos, tutorials, and samples to help you get the most from ASP. NET MVC. If you have any questions about ASP. NET MVC visit our forums.

其中，"http：//asp. net/mvc" 超链接到 http：//asp. net/mvc 的网站，"videos, tutorials, and samples" 添加浅蓝色背景，"our forums" 链接到 http：//forums. asp. net/1146. aspx/1？MVC。

下面的文字的内容为：

We suggest the following：

① Getting Started

ASP. NET MVC gives you a powerful, patterns-based way to build dynamic websites that enables a clean separation of concerns and that gives you full control over markup for enjoyable, agile development. ASP. NET MVC includes many features that enable fast, TDD-friendly development for creating sophisticated applications that use the latest web standards. Learn more...

② Add NuGet packages and jump-start your coding

NuGet makes it easy to install and update free libraries and tools. Learn more...

③ Find Web Hosting

You can easily find a web hosting company that offers the right mix of features and price for your applications. Learn more...

其中，第一个 Learn more... 链接到 http：//www. asp. net/mvc；第二个链接到 http：//www. nuget. org/；第三个链接到 http：//www. asp. net/hosting。

About 页面如图 1.36 所示。

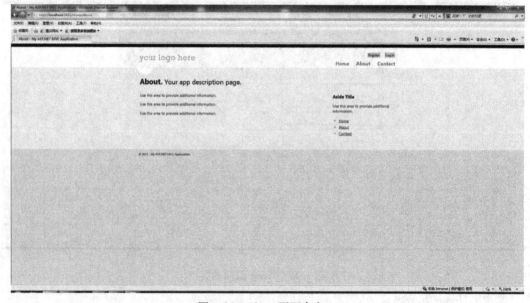

图 1.36　About 页面内容

Contact 页面内容如图 1.37 所示。

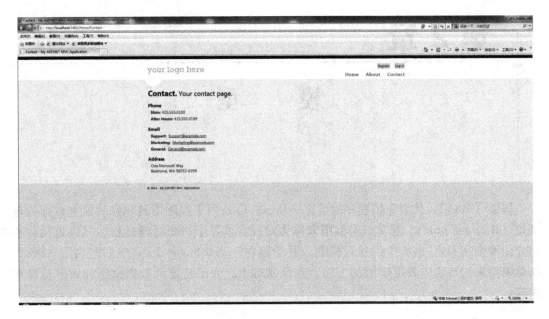

图 1.37　Contact 页面内容

第 2 章 模 型

模型（Model）代表应用程序的数据（Data）以及用于控制访问和修改这些数据的业务规则（Business Rule）。通常，模型用来作为对现实世界中一个处理过程的软件近似。本章将介绍模型在 MVC 框架中的执行机制，并给出在 Visual Studio 2012 中如何利用 ADO. NET 实体模型来构造实体数据模型的方法。在此基础上，介绍自定义数据模型和数据检验相关内容。

2.1 模型层概述与执行机制

模型（Model）层是 MVC 4 框架的最底层部分，它的职能主要有两个：①提供数据库的支持，包括对数据库中的数据进行增加（insert）、删除（delete）、修改（update）和查找（select）等具体操作；②提供软件模型、应用的具体实现过程，也就是最底层的方法的具体内容，封装给上层接口进行调用。对于以上两方面的内容，数据库的封装与支持是本章的主要内容，而对于模型方法的具体实现方面的内容，将在之后的章节中有具体体现，这里不再赘述。

在 MVC 框架中，模型层的执行机制如图 2.1 所示。

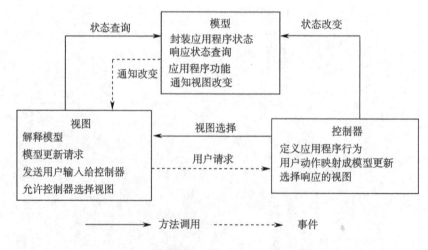

图 2.1 模型层在 MVC 框架中的执行机制

ASP. NET 的 MVC 4 框架中，模型表示应用程序的数据对象，以及相应的业务领域逻辑，包括数据验证和业务规则等内容。因此，模型是 ASP. NET MVC 4 框架的核心部分。

由于模型层应该只和数据及业务逻辑有关，也就是说，其不应该负责处理所有与数据或业务处理无关的操作，如不应该控制网站的执行流程等，因此，模型层应该只专注于如何有效地提供数据访问机制、交互环境、数据格式验证、业务逻辑验证等服务。

同时，由于模型的独立性非常高，因此，如果一个 Visual Studio 方案中有多个要开发的项目，应该将模型独立成一个项目，好让此模型项目能在不同的项目之间共享使用。

2.2 实体数据模型

MVC 4 中，实体数据类型是系统与数据库的交互接口，非常重要。不同于 Java 使用 JDBC-ODBC 的数据库连接方式，MVC 4 在数据库连接的基础之上，将数据库在系统中抽象成为一个具体的数据库实体对象，在实体对象中映射出数据库中的各个表、各个字段的具体内容，完全模拟数据库。也就是说，在编写的工程中，直接在系统中就可以查到各个数据表的组成、样式和取值等，不用再去数据库中进行查找。建立数据库实体的方法如例 2.1 所示。

【例 2.1】数据库实体的建立。

本例是对数据库实体建立步骤的具体说明。通过本例，读者可以完全掌握实体数据库建立的方法与机制。首先，数据库名称为 MyStudent，数据库具体结构见表 2.1～表 2.3。

表 2.1 Student（学生）表

序号	字段	字段描述	格式	长度	是否主码	备注
1	Sno	学号	nVarchar	5	是	
2	Sname	姓名	nVarchar	8		
3	Ssex	性别	nVarchar	2		
4	Sage	年龄	Numeric			
5	Sdept	所在系所	nVarchar	20		

表 2.2 Course（课程）表

序号	字段	字段描述	格式	长度	是否主码	备注
1	Cno	课程号	nVarchar	5	是	
2	Cname	课程名称	nVarchar	20		
3	Cpno	先行课号	nVarchar	5		外码
4	Ccredit	学分	Numeric			

表 2.3 SC（选课）表

序号	字段	字段描述	格式	长度	是否主码	备注
1	Sno	学号	nVarchar	5		
2	Cno	课程号	nVarchar	5		
3	Grade	成绩	Numeric			

数据库的初始赋值见表 2.4～表 2.6。

表 2.4　Student 表初始值

Sno	Sname	Ssex	Sage	Sdept
95001	李名	男	20	CS
95002	张明	女	19	IS
95003	刘红	女	19	MA
95004	陈列	男	19	IS

表 2.5　Course 表初始值

Cno	Cname	Cpno	Ccredit
1	数据库	5	4
2	数学		2
3	信息系统	1	4
4	操作系统	6	3
5	数据结构	7	4
6	数据处理		2
7	PASCAL 语言	6	4

表 2.6　SC 表初始值

Sno	Cno	Grade
95001	1	97
95001	2	90
95001	3	80
95002	2	86
95002	3	92

具体数据实体的建立过程见 1.4.2 节 Step01～Step05。建立好的数据库实体如图 2.2 所示。

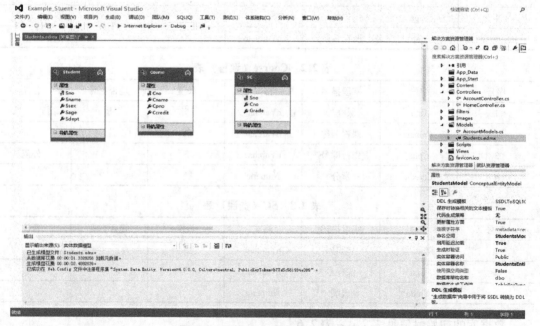

图 2.2　MyStudent 数据库实体图

需要注意的是，在Step05中勾选"确定所生成对象名称的单复数形式"复选框后，系统会将每个表以一个集合的复数形式给出，而每一个个体元素则以一个单一个体（单数）形式给出。也就是说，在创建实体数据类型时，实体数据类型是有单复数之分的，数据库和表的类型使用复数，实例使用单数，因此需要注意有无s的区分。

在数据实体中，右击实体表，在弹出的快捷菜单中选择"属性"，会在UI界面的右下方显示数据表的属性。通过属性，可以进行表名称、表对应集的修改，单击每一个表内容，如Student表的Sno属性，会在右下角显示Sno属性的相关属性。我们可以在这里进行表与属性的修改。

【例2.2】对表字段进行操作。

1）选择"Student"表，将表名称改为"PersonInSchool"，对应于数据库中的"Student"表。

2）选择"Student"表中的"Sname"字段，修改其最大长度为6，名称为"StudentName"，对应于数据库中的"Sname"属性，默认值设为"省略"；设置"Sage"的名称为"StudentAge"，对应于数据库中的"Sage"属性，设置默认值为"18"。

3）针对以上要求，在1.4.2节的例子中进行修改，完成以上内容，并保证系统的正确运行。

以上内容均在右侧属性栏中进行修改即可。需要注意的是，当完成以上修改后，程序中之前应用的字段都要进行相应的对照修改，这样才能保证系统的正常运行。

2.3 LINQ语句与使用

在MVC 4的数据部分，我们将不再使用普通的SQL语句作为数据库操作语言，而是使用语言集成查询（Language Integrated Query，LINQ）作为固定数据库的操作语言。

LINQ是ASP.NET中的新特性，它分为LINQ to Object、LINQ to SQL、LINQ to XML、LINQ to DataSet，并且它提供了统一的语法来查询多种异构数据源，开发人员可使用任何.NET语言（比如C#或者VB.NET）来查询数据源中的数据，而不用处理各种异构数据源之间的差异。LINQ允许开发人员查询任何实现了IEnumerable < > 接口的集合，如Array、List、XML、DOM或者SQL Server数据表。LINQ提供了翻译时的类型安全检查和动态类型组合等功能，这里重点讲解如何使用LINQ to SQL访问数据库。

2.3.1 使用LINQ的好处

LINQ与SQL相比，有不少的好处，特别是学过SQL的人员更容易上手。LINQ的优点主要表现在：

1）无须复杂的学习过程即可上手。
2）编写更少代码即可创建完整的应用。
3）更快开发错误更少的应用。
4）无须利用"奇怪"的编程技巧就可合并数据源。
5）提高新开发者的开发效率。
6）任何对象或数据源都可以定制实现LINQ适配器，为数据交互带来真正的便利。

2.3.2 LINQ to SQL 的预备知识

1. DataContext 类型介绍

DataContext 类型（数据上下文）是 System.Data.Linq 命名空间下的重要类型，用于把查询语句翻译成 SQL 语句，以及把数据从数据库返回给调用方和把实体的修改写入数据库。

DataContext 提供了以下一些实用的功能：

1）以日志形式记录 DataContext 生成的 SQL。
2）执行 SQL（包括查询和更新语句）。
3）创建和删除数据库。

DataContext 是实体和数据库之间的桥梁，首先我们需要定义映射到数据表的实体。

Step01：定义实体类。

```
1    Using System.Data.Linq.Mapping;
2    [Table(Name = "Student")]
3    public class Student
4    {
5        [Column(IsPrimaryKey = true)]
6        public string Sname {get; set;}
7        [Column(Name = "Sname")]
8        public string Ssex { get; set; }
9        [Column]
10       public int Sage {get; set;}
11       [Column]
12       public string Sdept {get; set;}
13   }
```

以 MyStudent 数据库为例，上述 Student 类被映射成一个表，对应数据库中的 Student 表。然后，在类型中定义了 4 个属性，对应表中的 4 个字段。其中，Sname 字段是主码，如果没有指定 Column 特性的 Name 属性，那么系统会把属性名作为数据表的字段名，也就是说，实体类的属性名就需要和数据表中的字段名一致。

Step02：现在，创建一个 ASP.NET 页面，然后在页面上加入一个 GridView 控件数据绑定。

Step03：使用 DataContext 类型把实体类和数据库中的数据进行关联。用户可以直接在 DataContext 构造方法中定义连接字符串，也可以使用 IDbConnection。

```
using System.Data.SqlClient;
IDbConnection conn = new SqlConnection("server = xxx;database = Northwind;uid = xxx;pwd = xxx");
DataContext ctx = new DataContext(conn);
```

Step04：然后，通过 GetTable 获取表示底层数据表的 Table 类型，显然，数据库中的 Student 表的实体是 Student 类型。随后的查询语句，即使用户不懂 SQL 应该也能看明白。从 Student 表中找出 Sname 以"李"开头的记录，并把 Sname、Ssex、Sage 及 Sdept 封装成新的匿名类型进行返回。

结果如图 2.3 所示。

学生姓名	性别	年龄	系所
李明	男	20	CS

图 2.3 运行结果

2. 匿名类型

在动态语言中，用户可能比较熟悉的概念是匿名类型（anonymous type）。当需要创建临时类型，却并不想创建类时，可以使用匿名类型。

下面的代码演示了如何在 C#中创建匿名类型。

```
var student = new {FirstName = "Stephen", LastName = "Walt"};
```

注意：student 变量并没有指定类型（JavaScript 或 VBScript 与之非常类似）。尽管如此，student 仍然具有它的类型，只是用户不知道它的名字而已，它是匿名的，理解这一点非常重要。

仅仅一行代码，我们即创建了一个新的类，又初始化了它的属性，十分简洁。在使用 LINQ to SQL 时，匿名类型非常有用，因为用户会发现经常需要创建一些临时功能的新类型。例如，当执行一个查询时，也许希望返回一个类，用来代表一些数据库列的集合，此时将需要创建一个代表这些列的临时类。

3. Lambda 表达式

```
1       var list = new [] { "aa", "bb", "ac" };
2       var result = Array.FindAll(list, s = > (s.IndexOf("a") > -1));
3       foreach (var v in result)
4       Console.WriteLine(v);
```

在上述这段代码中，添加了 C# 4.0 的新语法：s = >(s.IndexOf("a") > -1)，这就是一个 Lambda 表达式，现在只需要简单的一些字符就完成了原本需要单独创建一个方法的过程，而且语法相当简洁。

通过上面的例子，我们现在对 Lambda 表达式已经有了一个基本的了解。一个 Lambda 表达式通常由如下部分组成。

1) 首先是一个参数或者参数列表，也就是输入变量。在上述示例中，由于需要为委托传递一个字符串类型的变量，因此左侧的是 s 变量。

2) 接着是 = >符号，称为 Lambda 运算符，MSDN 中将这个符号读作"goes to"。

3) 最后是 Lambda 语句（块），可以是单条语句也可以是多个语句的语句块。为了加深印象，读者可以这样来理解 Lambda 表达式：

要处理的参数 = >处理这些参数的语句(块)

说明如下：

① 要处理的参数：可以有一个参数、多个参数，或者无参数。

② 处理这些参数的语句（块）：这部分就是我们平时编写函数时的实现部分（函数体）。

2.3.3 LINQ to SQL 的查询

使用 LINQ to SQL 进行查询，类似于创建一个反向的 SQL 查询。LINQ to SQL 查询以一个 from 子句开始，它指定了数据的位置，然后，指定 where 子句过滤数据，最后，指定用来表示数据的 select 子句（决定要返回的对象和属性）。

因此，下列查询：

```
1    var q =
2        from c in db.Customers
3        where c.City == "London"
4        select c;
```

将被 C#编译器翻译成下面的查询：

```
var q = db.Customers.Where(c => c.City == "London").Select(c => c);
```

第一个查询使用了查询语法，第二个查询使用了方法语法。这两种查询是相同的。

注意：使用方法语法的查询在 Where( ) 和 Select( ) 方法中允许使用 Lambda 表达式。Where( ) 方法中的 Lambda 表达式用来过滤数据，Select( ) 方法指明要返回的对象合适属性。

1. 简单查询

【例 2.3】筛选年龄为 20 的学生。

```
1    var q =
2        from e in db.Studnets
3        where e.Sage = 20
4        select e;
```

2. 条件查询

【例 2.4】使用 SELECT 和条件语句返回学生姓名和所在系所的序列，其中系所只能是 CS。

```
1    var q =
2        from p in db.Studnets
3        where p.Sdept = "CS"
4        select new
5        {
6            p.Sname,
7            sdept =
8            p.Sdept
9        };
```

【例 2.5】使用嵌套查询返回所有选课人的学号、姓名、课程号、课程名称、先行课号、学分和成绩。

```
1    var q =
2        from o in db.SCs
3        select new {
4            o.Sno,
5            Studnets =
6            from od in o.Student
```

7	where od.Sno = o.Sno
8	select od,
9	o.Cno,
10	Courses =
11	from oc in o.Course
12	where oc.Cno = o.Cno
13	select oc,
14	o.Grade,
15	};

【例2.6】使用 Count/Sum/Min/Max/Avg 操作符进行查询。

1) 查找最小年龄。

       var q = db.Studnets.Select(p = > p.Sage).Min();

2) 查找拥有最低学分的课程。

       var q = db.Course.Min(o = > o.Ccredit);

【例2.7】使用 OrderBy 按课程学分对课程进行排序。

1	var q =
2	from e in db.Courses
3	OrderBy e.Ccredit
4	select e;

【例2.8】使用 GroupBy 进行查询。

使用 OrderBy、Max 和 GroupBy 得到每个学生的最高成绩, 并按 Sno 对其进行排序。

1	var q =
2	from o in db.SCs
3	select new {
4	o.Sno,
5	Studnets =
6	from od in o.Student
7	where od.Sno = o.Sno
8	select od,
9	o.Cno,
10	Courses =
11	from oc in o.Course
12	where oc.Cno = o.Cno
13	select oc,
14	o.Grade,
15	};

3. 模糊查询

使用 LINQ to SQL 进行模糊查询, 可以有很多方法, 如 Length、Substring、Contain、StartsWith、IndexOf 等。

【例2.9】下面查询并返回以"李"开头的全部学生信息。

MyStudentsContext db = new MyStudentsDataContext();
var q = db.Students.where(m = > m.Sname.StartsWith("李"))

4. 多表查询

【例 2.10】 在 select 子句中使用外码来筛选成学生和成绩。

```
1    var q =
2    from e1 in db.Students
3    from e2 in e1.SCs
4    where e1.Sno == e2.Sno
5    select new {
6        Name = e1.Sname,
7        e2.Grade
8    };
```

像上程序 2、3 行所示，语句没有 join 和 into，则被翻译成 SelectMany；同时有 join 和 in 时，那么语句就被翻译为 GroupJoin。在这里，in 的概念是对其结果进行重新命名。

【例 2.11】 显式联接两个表并从这两个表投影出结果。

```
1    var q =
2    from c in db.Students
3    join o in db.SCs on c.Sno
4    equals o.Sno into orders
5    select new
6    {
7        c.Sname,
8        SelectClassNumber = orders.Count()
9    };
```

【例 2.12】 显式联接三个表并分别从每个表投影出结果。

```
1    var q =
2    from c in db.Students
3    join o in db.SCs on c.Sno
4    equals o.Sno into ords
5    join e in db.Courses on o.Cno
6    equals e.Cno into emps
7    select new
8    {
9        c.Sname,
10       c.Sage,
11       SelectClassNumber = orders.Count()
12   };
```

【例 2.13】 以 Students 作为主表，SCs 作为要连接的表，当 SCs 表中内容为空时，用 null 值填充。join 的结果重命名为 ords，使用 DefaultIfEmpty() 函数对其再次查询。其最后的结果中有个 Order，因为 "from o in ords.DefaultIfEmpty()" 是对 ords 组再一次遍历，所以，最后结果中的 Order 并不是一个集合。但是，如果没有 "from o in ords.DefaultIfEmpty()" 这条语句，而且最后的 select 语句写成 "select new { e.Sname, Order = ords }" 的话，那么 Order 就是一个集合。

```
1       var q =
2       from e in db.Students
3       join o in db.SCs on e.Sno equals o.Sno into ords
4       from o in ords.DefaultIfEmpty()
5       select new
6       {
7           e.Sname,
8           Order = o
9       };
```

2.3.4 LINQ to SQL 进行插入

插入记录时需要先创建一条新记录（对象），包括数据表中不能为 null 的所有字段，然后将这个对象作为属性成员添加到原有的类中，最后增加到数据库中。使用 LINQ to SQL 添加或插入新记录也是这样的步骤。首先，需要使用 InsertOnSubmit( ) 方法将一个实体添加到一个已经存在的表中。然后，调用 DataContext 中的 SubmitChanges( ) 对数据库执行 SQL INSERT 语句。

实例代码：

```
1   protected void Button3_Click(object sender, EventArgs e)
2   {
3       NorthwindDataContext db = new NorthwindDataContext();
4       Students pp = new Students { Sno = int.Parse(TextBox1.Text), Sname = Text-
5   Box4.Text };
6           //先生成一条新记录
7       db.Students.InsertOnSubmit(pp);
8           //将新记录加入到类中
9       db.SubmitChanges();
10          //将新记录增加到数据库中
11      }
12  }
```

2.3.5 LINQ to SQL 进行更新

使用 LINQ to SQL 进行更新时，可以通过修改实体属性并调用 DataContext 的 SubmitChanges( ) 方法来对 LINQ to SQL 实体和底层的数据库表进行更新。

实例代码：

```
1   protected void Button2_Click(object sender, EventArgs e)
2   {
3           NorthwindDataContext db = new NorthwindDataContext();
4   Students updateAge = db.Studnets.Single(m = >m.Sno == int.Parse("12"));
5           //先定位需要更新的记录
6   updateAge.Sage = decimal.Parse("12");
```

```
7                //指定字段修改的值
8            db.SubmitChanges();
9                //返回数据库修改
10       }
```

总结一下，这段代码就是先得到产品 ID = 12 的产品，然后修改产品的单价属性，最后调用 SubmitChanges() 方法向数据库提交修改。

2.3.6 LINQ to SQL 进行删除

使用 LINQ to SQL 进行删除时，可以采用下面的代码。

实例代码：

```
1       protected void Button4_Click(object sender, EventArgs e)
2       {
3           Students DD = db.Students.Single(m => m.Sno == int.Parse(TextBox8.Text));
4           db.Students.DeleteOnSubmit(DD);
5           db.SubmitChanges();
6       }
```

该段代码一开始从 Student 数据表中获取 m.Sno == int.Parse(TextBox8.Text) 的记录，然后用 DeleteOnSubmit() 的方法将该条记录删除，最后调用 SubmitChanges() 方法向数据库提交修改。

2.4 模型的数据校验

对用户输入的验证以及强制业务规则/逻辑是大多数 Web 应用的核心需求。ASP.NET MVC 4 包含了许多新的特性，明显简化了对用户输入的验证以及在模型/视图模型中对验证逻辑的强行实施。这些特性是这样设计的，验证逻辑在服务器上执行，也可以选择在客户端通过 JavaScript 来执行。ASP.NET MVC 4 中的验证设施和特性这样设计，以便：

1）开发人员可以轻松地利用内置于 .NET 框架中的 DataAnnotation 验证支持。DataAnnotation 提供了一个非常简便的方式，可以使用最少的代码在对象和属性上用声明的方式添加验证规则。

2）开发人员可以集成他们自己的验证引擎，或者利用现有的验证框架，如 Castle 验证器或 EntLib 验证库。ASP.NET MVC 4 的验证特性是设计用来在利用新的 ASP.NET MVC 4 的验证设施（包括客户端验证、模型绑定验证等）的同时，简化任何类型的验证架构的插入的。

这意味着在常见的应用场景中启用验证是极其容易的，同时对更高级的场景还能保持极高的灵活性。

2.4.1 非数据库类 DataAnnotation 启用验证

下面在 ASP.NET MVC 4 中全程演示一个简单的 CRUD 场景示例，利用新的内置 DataAnnotation 验证支持。具体来说，就是实现一个 "Create" 表单来允许用户输入个人数据。

我们想要确保在输入数据保存到数据库之前是合法的，如果不合法，就显示合适的错误消息。

我们想要使得这个验证同时在服务器端和客户端（通过 JavaScript）发生。我们还想要确保我们的代码遵循 DRY 原则（Don't Repeat Yourself，即不重复自己），这意味着我们应该只在一处实施验证规则，然后使得我们的控制器、action 方法和视图来"兑现"这个承诺。

接下来将使用 Visual Studio 2012，并利用 ASP. NET MVC 4 来实现上述的场景。用户也可以使用 Visual Studio 2010 及 ASP. NET MVC 4 来实现完全一样的场景。

Step01：实现 FriendsController（没有校验）。

首先在一个新的 ASP. NET MVC 4 项目中添加一个简单的"Person"类，像下面这样：

```
1    Public class Person
2    {
3        Public string FirstName { set; get; }
4        Public string LastName { set; get; }
5        Public int Age { set; get; }
6        Public String Email { set; get; }
7    }
```

然后在项目中添加一个"FriendsController"控制器类，添加两个"Create" action 方法。第一个 action 方法是在对 /Friends/Create URL 的 HTTP-GET 请求进来时调用的，它会显示一个空白的表单，用来输入个人数据；第二个 action 方法是在对 /Friends/Create URL 的 HTTP-POST 请求进来时调用的，它会将提交的表单输入映射到一个 Person 对象，确认没有绑定错误发生，如果是合法的，最终会将数据保存到数据库中（在下文我们会实现相关的数据库工作）。如果提交的表单输入是不合法的，那么该 action 方法会重新显示带有错误的表单。

```
1    Namespace test1.Controllers
2    {
3        Public ActionResult Create()
4        {
5            Person newFriend = new Person();
6            Return View(newFriend);
7        }
8        [HttpPost]
9        Public ActionResult Create(Person friendToCreate)
10       {
11           If(ModelState.IsValid)
12           {
13               //如果模型的验证成功,则在此进行数据持久层的工作
14               Return Redirect("/");
15           }
16           //如果模型的验证未成功,则将用户填写数据重新回填,并提示错误
17           return View(friendToCreate);
18       }
```

在实现了控制器之后，可以在 Visual Studio 中的其中一个 action 方法中右击，在弹出的快捷菜单中选择"添加视图"命令，这会弹出"添加视图"对话框，选择自动生成传入对

象为 Person 的"Create"视图，如图 2.4 所示。

图 2.4 构建视图

然后，Visual Studio 会在我们项目的 \ Views \ Friends \ 目录中生成一个含有框架代码的 Create.aspx 视图文件。注意下面这段代码，它利用了 ASP. NET MVC 4 中新的强类型 HTML 辅助方法。

```
1    <% using (Html.BeginForm()) {% >
2    <% : Html.ValidationSummary(true)% >
3    <fieldset>
4    <legend>Fields</legend>
5    <div class = "editor-label">
6    <% : Html.LabelFor(model = >model.FirstName) % >
7    <div>
8    <div class = "editor-field">
9    <% : Html.TextBoxFor(model = >model.FirstName) % >
10   <% : Html.ValidationMessageFor(model = >model.FirstName) % >
11   <div>
12   <div class = "editor-label">
13   <% : Html.LabelFor(model = >model.Age) % >
14   <div>
15   <div class = "editor-field">
16   <% : Html.TextBoxFor(model = >model.Age) % >
17   <% : Html.ValidationMessageFor(model = >model.Age) % >
18   <div>
19   <div class = "editor-label">
20   <% : Html.LabelFor(model = >model.Email) % >
```

```
21          <div>
22            <div class = "editor-field">
23              <%: Html.TextBoxFor(model = >model.Email) %>
24              <%: Html.ValidationMessageFor(model = >model.Email) %>
25            <div>
26          <p>
27            <input type = "submit " value = "Create" />
28          </p>
29        </fieldset>
```

此时，当我们运行该应用，并且访问/Friends/Create URL 时，将得到一个可以输入数据的空白表单，如图 2.5 所示。

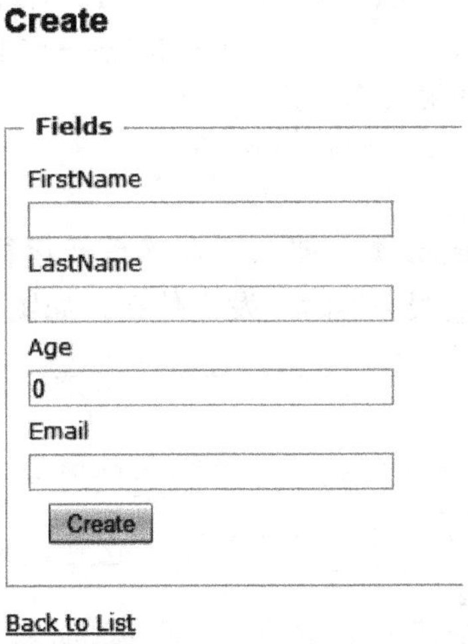

图 2.5 Create.aspx 视图

但是，由于我们还没有在应用中实现任何验证，因此无法阻止用户在表单中输入不合法的内容，然后将其提交到服务器中。

Step02：使用 DataAnnotation 来启用验证。

现在，让我们来更新应用，执行一些基本的输入验证规则。我们将在 Person 模型对象上实现这些规则，而不是在控制器或视图中实现。在 Person 对象上实现这些规则的好处：这将确保这些验证在应用中任何使用 Person 对象的场景中都会执行（假如后来添加了编辑场景的话），从而避免在多处重复这些规则。

ASP.NET MVC 4 允许开发人员轻松地在模型或视图模型类上添加声明式验证，然后 ASP.NET MVC 在应用中实施模型绑定操作时，这些验证规则就会自动执行。举例来说，让我们更新 Person 类，在其中添加几个验证特性，这样做的话，将在文件的顶部添加一条对"System.Com ponentModel.DataAnnotations"命名空间的"using"语句，然后在 Person 的属性上使用[Required]、[StringLength]、[Range] 和 [RegularExpression] 验证特性（这几个特性都是在那个命名空间中实现的）。

```
1    public class Person
2    {
3        [Required(ErrorMessage = "您的姓氏必须填写")]
4        [StringLength(5, ErrorMessage = "姓氏最大长度不能超过5个字符")]
5        public string FirstName { set; get;}
6        [Required(ErrorMessage = "您的名字必须填写")]
7        [StringLength(5, ErrorMessage = "名字最大长度不能超过5个字符")]
8        public string LastName { set; get;}
9        [Required(ErrorMessage = "年龄必须填写")]
10       [Range(0, 200, ErrorMessage = "年龄应在0-200之间")]
11       public int Age { set; get;}
12       [Required(ErrorMessage = "Email地址必须填写")]
13       [RegularExpression("'[a-z]([a-z0-9]*[-_]?[a-z0-9]+)*@
14   ([a-z0-9]*[-_]?[a-z0-9]+)+[.][a-z]{2,3}([.][a-z]{2})?$'", Er-
15   rorMessage = "Email地址错误")]
16       public string FirstName { set; get;}
17   }
```

既然添加了验证特性到 Person 类，那么现在来重新运行我们的应用，此时需要查看在输入不合法输入项并将其提交回服务器时会发生什么，如图 2.6 所示。

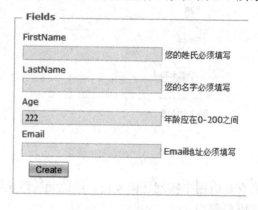

图 2.6　验证效果

从图 2.6 我们可以看出，此例是一个不错的校验体验，带不合法输入的文本元素以红色高亮显示，我们指定的验证错误消息也显示给了用户。另外，表单还保留用户原先输入的数据，这样他们不用重新填写什么。上述内容究竟如何实现呢？

要理解这个行为，让我们看一下处理表单的 POST 场景的 Create action 方法。

```
1    [HttpPost]
2    public ActionResult Create(Person friendToCreate)
3    {
4        if(ModelState.IsValid)
5        {
6            //如果模型的验证成功，则在此进行数据持久层的工作
7        return Redirect("/");
```

```
8          }
9          //如果模型的验证未成功，则将用户填写数据重新回填，并提示错误
10         return View(friendToCreate);
11     }
```

在我们的 HTML 表单提交回服务器时，上面的方法就会被调用。因为该 action 方法接收一个 "Person" 对象为参数，所以 ASP. NET MVC 会创建一个 Person 对象，自动将传输进来的表单输入数值映射到该对象上。作为该过程的一部分，ASP. NET MVC 还会检查该 Person 对象上的 DataAnnotation 验证特性是否合法。如果一切都合法，那么我们代码中的 ModelState. IsValid 检查就会返回 "true"，在这种情形下，我们（最终）将把该 Person 对象保存到数据库中，然后重定向到主页。

如果 Person 对象上有任何验证错误，那么 action 方法就会以该不合法 Person 对象的数据重新显示表单，这是通过上面代码片段中最后一条代码语句实现的。

然后，错误消息就会显示在我们的视图中，因为 Create 表单在每一个 <%: Html. TextBoxFor( ) %> 辅助方法的调用旁都有一个 <%: Html. ValidationMessageFor( ) %> 辅助方法调用。Html. ValidationMessageFor( ) 辅助方法会针对传入视图的任何不合法的模型属性输出合适的错误消息。

```
1      < div class = "editor-label" >
2      < % : Html. LabelFor(model = >model. FirstName) % >
3      < div >
4      < div class = "editor-field" >
5      < % : Html. TextBoxFor(model = >model. FirstName) % >
6      < % : Html. ValidationMessageFor(model = >model. FirstName) % >
7      < div >
```

这个模式/方式有一个好处，就是非常容易配置。另外，它还允许我们轻松地添加或改变 Person 类上的验证规则，而不必改变控制器或视图中的任何代码。这个在一个地方指定验证规则，然后在所有的地方都会被承诺和遵守的能力，允许我们以最小的努力快速地发展我们的应用和规则。

Step03：启用客户端验证。

目前我们的应用只能做服务器端的验证，这意味着我们的终端用户需要将表单提交到服务器才能看到任何验证错误消息。

ASP. NET MVC 4 的验证架构中有一个非常不错的功能，那就是它同时支持服务器端和客户端验证。如果要启用这个功能，那么我们要做的就是在视图中添加两个 JavaScript 引用，即编写下面的代码。

```
1      < h2 >Create </h2 >
2      <script src ="../../Scripts/MicrosoftAjax.js" type ="text/javascript"></script>
3      <script src ="../../Scripts/MicrosoftAjax.js" type ="text/javascript"></script>
```

在添加了上述代码语句后，ASP. NET MVC 4 就会使用我们添加到 Person 类上的验证元数据，为我们连接好客户端 JavaScript 验证逻辑。这意味着，当用户使用 <Tab> 键跳出一个不合法的输入元素时，就会得到瞬时的验证错误，如图 2.7 所示。

让我们重新运行应用，在 "FirstName"、"LastName" 和 "Email" 文本框中输入合法的

数值，然后尝试单击"Create"按钮，发现会有错误提示，如图2.8所示。注意，其实我们不必访问服务器就会得到超出年龄范围值的瞬时错误消息。

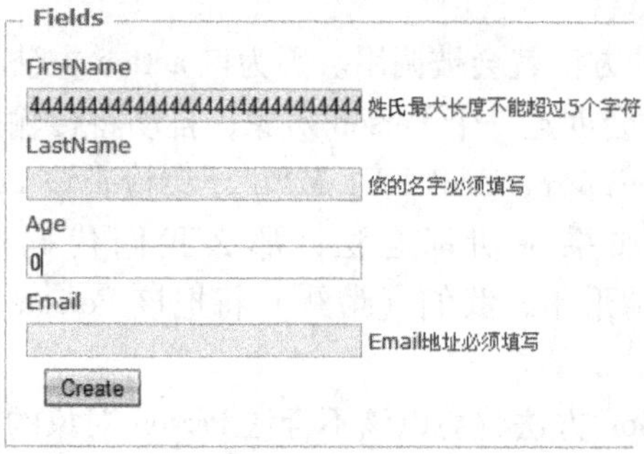

图2.7　瞬时的验证错误 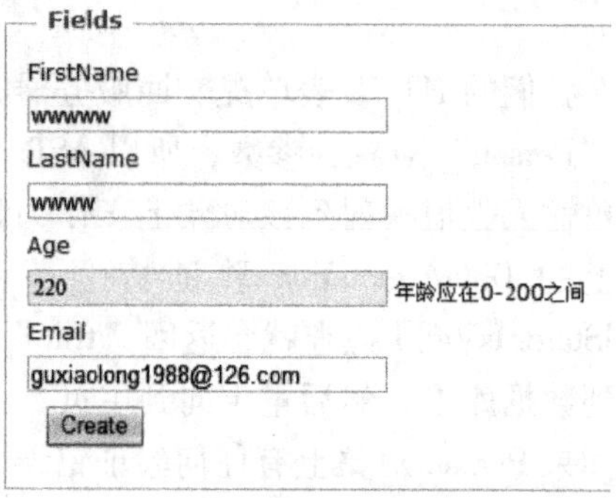　　　　　　　　　图2.8　提交后的验证错误

这项技术带来的好处是，我们不必编写自己的任何JavaScript就能启用上面的验证逻辑。我们可以在一个地方指定规则，然后在整个应用中得到执行，即同时在客户端和服务器端。

值得读者注意的是，由于安全的原因，服务器端验证规则总是执行的，即使用户启用了客户端支持。这是为了避免黑客尝试绕过客户端规则，"哄骗"攻击（spoof）用户的服务器。

ASP.NET MVC 4中的客户端JavaScript验证支持可与用户在ASP.NET MVC应用中使用的任何验证框架/引擎协作，它并不要求用户使用DataAnnotation验证方式，所有的基础设施是独立于DataAnnotation的，可以与Castle验证器、EntLib验证应用块，或者用户选择的任何定制验证方案协作使用。

如果用户不想使用我们的客户端JavaScript文件，那么也可以将其替换成jQuery验证插件，而使用那个库。ASP.NET MVC Futures的下载包还包括针对ASP.NET MVC 4服务器端验证框架启用jQuery验证的支持。

Step04：创建自定义的[Email]验证特性。

.NET框架中的System.ComponentModel.DataAnnotations命名空间包括了众多内置验证特性。在上面的例子中，使用了其中的4个：[Required]、[StringLength]、[Range]和[RegularExpression]。

用户也可以定义自己的定制验证特性，然后应用它们。用户可以通过继承自System.ComponentModel.DataAnnotations命名空间中的ValidationAttribute基类，定义完全定制的特性。如果用户只想扩展它们的基本功能，那么也可以选择继承任何现有的验证特性。

例如，为了帮助清理Person类中的代码，我们也许想要创建一个新的[Email]验证特性，将检查合法Email的正则表达式封装起来。如果想要这样做的话，那么只要像下列代码那样继承RegularExpressionAttribute基类，然后用合适的Email正则表达式调用RegularExpressionAttribute基类的构造器即可。

```
1    using System;
2    using System.Collections.Generic;
3    using System.Linq;
```

```
4     using System.Web;
5     using System.ComponentModel.DataAnnotations;
6     namespace test1.Models.validate
7     {
8         public class EmailAttribute : RegularExpressionAttribute
9         {
10            public EmailAttribute():base("'[a-z]([a-z0-9]*[-_]?[a-z0-9]+)*@([a
11    -z0-9]*[-_]?[a-z0-9]+)+[.][a-z]{2,3}([.][a-z]{2})?$'"){}
12        }
13    }
```

然后将 Person 类更新成使用新的 [Email] 验证属性，换掉先前使用的正则表达式，这样会使得我们的代码更简洁，封装效果也更好。

```
1     public class Person
2     {
3         [Required(ErrorMessage = "您的姓氏必须填写")]
4         [StringLength(5, ErrorMessage = "姓氏最大长度不能超过 5 个字符")]
5         public string FirstName { set; get;}
6         [Required(ErrorMessage = "您的名字必须填写")]
7         [StringLength(5, ErrorMessage = "名字最大长度不能超过 5 个字符")]
8         public string LastName { set; get;}
9         [Required(ErrorMessage = "年龄必须填写")]
10        [Range(0, 200, ErrorMessage = "年龄应在 0 - 200 之间")]
11        public int Age { set; get;}
12        [Required(ErrorMessage = "Email 地址必须填写")]
13        [Email(ErrorMessage = "Email 地址错误")]
14        public string FirstName { set; get;}
15    }
```

在建立定制的验证特性时，用户还可以在服务器端及客户端指定通过 JavaScript 执行的验证逻辑。

除了建立可用于对象上个别属性的验证特性外，还可以将验证特性用于类的层次，这允许对一个对象中的多个属性实施验证逻辑。若要查看相关案例，可以参阅包含在默认 ASP.NET MVC 4 应用项目模板 AccountModels.cs/vb 文件中的 "PropertiesMustMatchAttribute" 定制特性（在 Visual Studio 2010 中，选择"文件"→"新 ASP.NET MVC 4 Web 项目"，然后查询该类）。

Step05：逻辑层的需求校验。

校验中除了基本的长度、必填、范围等之外，还有符合逻辑要求的校验。这些校验除了用上面的自定义方法之外，还可以在控制层进行相关的检验。

```
1     [HttpPost]
2     public ActionResult Create(Person friendToCreate)
3     {
4         if(ModelState.IsValid)
```

```
5        {
6            //在此可以进行逻辑校验
7            //如果模型的验证成功，则在此进行数据持久层的工作
8            return Redirect("/");
9        }
10       //如果模型的验证未成功，则将用户填写数据重新回填，并提示错误
11       return View(friendToCreate);
12   }
```

在上述代码段的相应位置可以自己编写相关的方法进行校验。

Step06：逻辑层的需求校验。

现在让我们实现将朋友数据保存到数据库所需的逻辑。

至此，我们只用了简单的（plain-old）C#类（有时称为"POCO"类，即"plain old CLR（or C#）object"）。我们可以使用的一个方案是，编写一些单独的持久代码，将已经编写好的现有类映射到数据库。目前，像 NHibernate 这样的对象关系映射（Object Relational Mapping，ORM）方案已经可以非常好地支持 POCO/PI 这样风格的映射。随 .NET 4 发布的 ADO.NET 实体框架（Entity Framework，EF）也支持 POCO / PI 映射，而且就像 NHibernate 那样，EF 也能使用以"只用代码"（code only）的方式（没有映射文件，也不需要设计器）定义持久性映射的功能。

如果我们的 Person 对象以这种方式映射到数据库，那么不用对 Person 类做任何改动，也不用改动任何验证规则，它还会继续完好地进行工作。

但是，如果我们想要使用图形工具来进行 ORM 映射，那该怎么办呢？

2.4.2 数据库类 DataAnnotation 启用验证

目前，使用 Visual Studio 的许多开发人员并不编写他们自己的 ORM 映射/持久逻辑，而是使用 Visual Studio 中内置的设计器来帮助管理这样的映射逻辑。

使用 DataAnnotation（或者任何其他形式的基于特性的验证）时经常被问到的问题是："如果你手头的模型对象是由 GUI 设计器创建/维护的话，那么该如何使用这些特性？"例如，与我们一直在使用的 POCO 风格的 Person 类不同，而是在 Visual Studio 中通过像 LINQ to SQL 或 ADO.NET EF 设计器这样的 GUI 映射工具定义/维护我们的 Person 类的话，该怎么办呢？

图 2.9 展示了在 Visual Studio 2010 中使用 ADO.NET EF 设计器定义的一个 Person 类。上方的窗口定义了 Person 类，下方的窗口展示了该类的属性是如何映射到数据库中"People"表的映射编辑器的。当用户在设计器上单击保存时，它会自动在项目中生成一个 Person 类。这虽然很棒，但每次用户做了改动，并单击保存时，它就会重新生成 Person 类，这会导致在对象上声明的任何验证特性的丢失。

将额外的基于特性的元数据（如验证特性）施加到由 Visual Studio 设计器自动生成/维护的类的一个方法是，采用一个我们称为"伙伴类"（buddy classes）的技术。基本来说，用户创建另外一个类，包含用户的验证特性和元数据，然后通过将"MetadataType"特性施加到一个与工具生成的类一起编译的 Partial 类上，将其与由设计器生成的类连接起来。例

图 2.9　数据库映射文件

如，如果我们想要将前面用到的验证规则施加到由 LINQ to SQL 或 ADO.NET EF 设计器维护的 Person 类上，那么可以更新我们的验证代码，使其存在于一个单独的 "Person_Validation" 类上，使用像下面这样的代码将其连接到由 Visual Studio 创建的 "Person" 类上。

```
1   [MetadataType(typeof(Person_Validation))]
2   public partial class Person
3   {
4   }
5   [Bind(Exclude = "id")]
6   public class Person_Validation
7   {
8       [Required(ErrorMessage = "您的姓氏必须填写")]
9       [StringLength(5, ErrorMessage = "姓氏最大长度不能超过 5 个字符")]
10      public string FirstName { set; get;}
11      [Required(ErrorMessage = "您的名字必须填写")]
12      [StringLength(5, ErrorMessage = "名字最大长度不能超过 5 个字符")]
13      public string LastName { set; get;}
14      [Required(ErrorMessage = "年龄必须填写")]
15      [Range(0, 200, ErrorMessage = "年龄应在 0 - 200 之间")]
16      public int Age { set; get;}
17      [Required(ErrorMessage = "Email 地址必须填写")]
18      [Email(ErrorMessage = "Email 地址错误")]
19      public string FirstName { set; get;}
20  }
```

上面的做法虽然没有纯粹的 POCO 方法那么优雅，但其好处是，可以用于 Visual Studio 中任何工具或设计器生成的代码。

最后一步，无论是否采用了 POCO 或工具生成的 Person 类，都需将合法的数据保存到数据库中。

这只要求我们用三行代码将 FriendsController 类中的注释替换掉，这三行代码将新朋友信息保存到数据库。下面是 FriendsController 类的完整代码（使用了 ADO. NET EF 进行数据库持久化工作）。

```
1        [HttpPost]
2        public ActionResult Create(Person friendToCreate)
3        {
4            if(ModelState.IsValid)
5            {
6                //在此可以进行逻辑校验
7                //如果模型的验证成功，则在此进行数据持久层的工作
8                personEntities p = new personEntities();
9                p.AddToPerson(friendToCreate);
10               p.SaveChanges();
11               return Redirect("/");
12           }
13           //如果模型的验证未成功，则将用户填写数据重新回填，并提示错误
14           return View(friendToCreate);
15       }
```

现在，当我们访问 /Friends/Create URL 时，可以轻松地添加新数据到我们的数据库中。

对于所有数据的验证，都是同时在客户端和服务器端进行的。我们可以轻易地在一个地方添加/修改/删除验证规则，而由整个应用中的所有的控制器和视图来执行这些规则。

※ 习 题

1. 在 MVC 中，模型的作用是什么？在模型中如何保证数据库中数据的安全性？
2. 使用 LINQ 语句完成以下操作：
1) 不使用视图，联合三张表，输入表的内容，并按照其中的两个属性进行排序。
2) 在模型中自定义数据结构，并将此数据库嵌入到 LINQ 语句中使用。
3. 对于例 2.1 中的表及数据库，完成任意两个表的级联显示，完成对表的增加、删除、修改和查询的操作，并完成操作时的数据校验工作。

※ 综合应用

本书将在第 2、3 和 5 章的综合应用中建立一个拥有基本的购物车功能的网站。其中，此处模块中主要讲解数据库和模型层的编写。

本例中的数据库主要包括 ProductCategories、Products、Members、Orders 和 OrderDetailItems 5

张表。

模型层需要建立 8 个类文件，分别为：Cart.cs、Member.cs、MemberLoginViewModel.cs、MvcShoppingContext.cs、OrderDetail.cs、OrderHeader.cs、Product.cs 和 ProductCategory.cs。

Cart.cs 记录了选购的商品和数量，代码为：

```
1       [DisplayName("选购商品")]
2       [Required]
3       public Product Product { get; set; }
4       [DisplayName("选购数量")]
5       [Required]
6       public int Amount { get;set; }
```

Member.cs 记录会员的注册信息。注册信息包括：会员账号、会员密码、中文姓名、网络昵称、会员注册时间、会员启用确认码。示例代码如下。

```
1    [DisplayName("会员资料")]
2        [DisplayColumn("Name")]
3        public class Member
4        {
5            [Key]
6            public int Id { get; set; }
7            [DisplayName("会员账号")]
8            [Required(ErrorMessage = "请输入 Email 地址")]
9            [Description("我们直接以 Email 当成会员的登录账号")]
10           [MaxLength(250, ErrorMessage = "Email 地址长度无法超过 250 个字符")]
11           [DataType(DataType.EmailAddress)]
12           [Remote("CheckDup", "Member", HttpMethod = "POST", ErrorMessage =
13    "您输入的 Email 已经有人注册过了!")]
14           public string Email { get; set; }
15           [DisplayName("会员密码")]
16           [Required(ErrorMessage = "请输入密码")]
17           [MaxLength(40, ErrorMessage = "请输入密码")]
18           [Description("密码将以 SHA1 进行散列运算,通过 SHA1 散列运算后的结果转为
19    HEX 表示法的字符串长度皆为 40 个字符")]
20           [DataType(DataType.Password)]
21           public string Password { get; set; }
22           [DisplayName("中文姓名")]
23           [Required(ErrorMessage = "请输入中文姓名")]
24           [MaxLength(5, ErrorMessage = "中文姓名不可超过 5 个字")]
25           [Description("暂不考虑用英文注册会员的情况")]
26           public string Name { get; set; }
27    …
28           public virtual ICollection<OrderHeader> Orders { get; set; }
29       }
```

MemberLoginViewModel.cs 是用户登录使用的类，记录了会员账号和密码。示例代码

如下。

```csharp
1     public class MemberLoginViewModel
2     {
3         [DisplayName("会员账号")]
4         [DataType(DataType.EmailAddress, ErrorMessage = "请输入您的 Email 地址")]
5         
6         [Required(ErrorMessage = "请输入{0}")]
7         public string email { get; set; }
8         [DisplayName("会员密码")]
9         [DataType(DataType.Password)]
10        [Required(ErrorMessage = "请输入{0}")]
11        public string password { get; set; }
12    }
```

OrderDetail.cs 是订单类，记录了订单明细，包括订单表头、订购商品，商品售价和购买数量等信息。示例代码如下。

```csharp
1     [DisplayName("订单明细")]
2     [DisplayColumn("Name")]
3     public class OrderDetail
4     {
5         [Key]
6         public int Id { get; set; }
7         [DisplayName("订单表头")]
8         [Required]
9         public virtual OrderHeader OrderHeader { get; set; }
10        [DisplayName("订购商品")]
11        [Required]
12        public Product Product { get; set; }
13        [DisplayName("商品售价")]
14        [Required(ErrorMessage = "请输入商品售价")]
15        [Range(99, 10000, ErrorMessage = "商品售价必须介于 99~10,000 之间")]
16        [Description("由于商品售价可能会经常变动,因此必须保留购买时的商品售价")]
17        [DataType(DataType.Currency)]
18        public int Price { get; set; }
19        [DisplayName("购买数量")]
20        [Required]
21        public int Amount { get; set; }
22    }
```

OrderHeader.cs 记录了用户的订单信息，包括订购会员名称、收件人姓名、联系电话、派送地址和金额等信息。示例代码如下。

```csharp
1     [DisplayName("订单表头")]
2     [DisplayColumn("DisplayName")]
3     public class OrderHeader
```

```csharp
4       {
5           [Key]
6           public int Id { get; set; }
7           [DisplayName("订购会员")]
8           [Required]
9           public virtual Member Member { get; set; }
10          [DisplayName("收件人姓名")]
11          [Required(ErrorMessage = "请输入收件人姓名,例如:小明")]
12          [MaxLength(40, ErrorMessage = "收件人姓名长度不可超过 40 个字符")]
13          [Description("订购的会员不一定就是收到商品的人")]
14          public string ContactName { get; set; }
15          [DisplayName("联系电话")]
16          [Required(ErrorMessage = "请输入您的联系电话,例如:12345678")]
17          [MaxLength(25, ErrorMessage = "电话号码长度不可超过 25 个字符")]
18          [DataType(DataType.PhoneNumber)]
19          public string ContactPhoneNo { get; set; }
20          [DisplayName("派送地址")]
21          [Required(ErrorMessage = "请输入商品派送地址")]
22          public string ContactAddress { get; set; }
23      …
24          [NotMapped]
25          public string DisplayName
26          {
27              get { return this.Member.Name + "于" + this.BuyOn + "订购的商品"; }
28          }
29          public virtual ICollection<OrderDetail>OrderDetailItems { get; set; }
30      }
```

Product.cs 用于记录商品的上架信息,包括:商品名称、商品类别和上架时间等信息。示例代码如下。

```csharp
1   [DisplayName("商品信息")]
2   [DisplayColumn("Name")]
3   public class Product
4   {
5       [Key]
6       public int Id { get; set; }
7       [DisplayName("商品类别")]
8       [Required]
9       public virtual ProductCategory ProductCategory { get; set; }
10      [DisplayName("商品名称")]
11      [Required(ErrorMessage = "请输入商品名称")]
12      [MaxLength(60, ErrorMessage = "商品名称不可超过 60 个字")]
13      public string Name { get; set; }
14  …
```

```
15          [DisplayName("上架时间")]
16          [Description("如果不设定上架时间,代表此商品永不上架")]
17          public DateTime PublishOn { get; set; }
18      }
```

ProductCategory.cs 用于记录商品的类别名称。示例代码如下。

```
1       [DisplayName("商品类别")]
2       [DisplayColumn("Name")]
3       public class ProductCategory
4       {
5           [Key]
6           public int Id { get; set; }
7           [DisplayName("商品类别名称")]
8           [Required(ErrorMessage = "请输入商品类别名称")]
9           [MaxLength(20, ErrorMessage = "类别名称不可超过20个字")]
10          public string Name { get; set; }
11          public virtual ICollection<Product> Products { get; set; }
12      }
```

MvcShoppingContext.cs 用于建立类与数据库的对应关系,将之前建立的 Model 类对应到数据库中。示例代码如下。

```
1       public MvcShoppingContext()
2               : base("name=DefaultConnection")
3       {
4       }
5       public DbSet<ProductCategory> ProductCategories { get; set; }
6       public DbSet<Product> Products { get; set; }
7       public DbSet<Member> Members { get; set; }
8       public DbSet<OrderHeader> Orders { get; set; }
9       public DbSet<OrderDetail> OrderDetailItems { get; set; }
10      }
```

其中,这个类引用了 EntityFramework.dll 的 DbContext 作为基类来使用。

建立好以上类后,在 Controllers 目录下建立 MemberController.cs,在其中输入 Register 代码,如下所示。

```
1       public class MemberController : BaseController
2       {
3           // 会员注册页面
4           public ActionResult Register()
5           {
6               return View();
7           }
8           //密码哈希所需的 Salt 随机数值
9           private string pwSalt = "A1rySq1oPe2Mh784QQwG6jRAfkdPpDa90J0i";
10          //写入会员资料
```

```csharp
11            [HttpPost]
12            public ActionResult Register([Bind(Exclude = "RegisterOn,AuthCode")] Member member)
13
14            {
15                //检查会员是否已存在
16                var chk_member = db.Members.Where(p => p.Email == member.Email).FirstOrDefault();
17
18                if (chk_member != null)
19                {
20                    ModelState.AddModelError("Email", "您输入的 Email 已经有人注册过了!");
21
22                }
23                if (ModelState.IsValid)
24                {
25                    //将密码加密之后进行哈希运算以提升会员密码的安全性
26                    member.Password = FormsAuthentication.HashPasswordForStoringInConfigFile(pwSalt + member.Password, "SHA1");
27
28                    //会员注册时间
29                    member.RegisterOn = DateTime.Now;
30                    //会员验证码,采用 Guid 当成验证码内容,避免有会员使用重复的验证码
31                    member.AuthCode = null;
32                    db.Members.Add(member);
33                    db.SaveChanges();
34                    return RedirectToAction("Index", "Home");
35                }
36                else
37                {
38                    return View();
39                }
40            }
```

然后，在 View 层中建立对应的页面 Register.cshtml 来调用这个注册代码完成注册功能。

第 3 章

控 制 器

控制器负责控制用户与 MVC 应用程序的交互方式。控制器决定在用户发出浏览器请求时向用户发送什么样的响应。控制器只是一个类（如 Visual Basic 或 C# 类），它是 ASP. NET MVC 的核心。但 Controller 并不负责决定内容应该如何显示，仅仅将特定形态的内容响应给 ASP. NET MVC 架构，最后才由 ASP. NET MVC 架构依据响应的形态来决定如何将内容响应给浏览器。

3.1 控制器概述

控制器层（即 Controller 层）主要有两大功能（见图 3.1）：

1) 用户通过视图层（将在第 5 章介绍）输入有关数据后，控制器层将接收到相应数据，通过有关代码的加工处理后，传递给模型层，进行相应业务逻辑的处理和相应数据的存取。

2) 对从模型层获取的原始数据进行加工处理，转换成视图层能直接使用的数据后，传递给视图层，由视图层负责和用户之间的交互。

总的来说，控制器层是模型层和视图层之间沟通的桥梁。

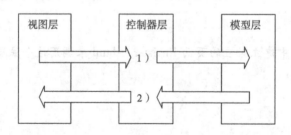

图 3.1 控制器层功能示意图

在 MVC 4 中，模型层中的函数和视图层中的视图，一般情况下是可以复用的。而控制器层的有关代码就相当于映射，使相应的函数和视图之间能建立起联系，以实现相应功能。这种映射是通过控制器中的动作（即 Action）来实现的（见图 3.2）。每个 Action 只能为一个视图服务，但可以调用多个函数；而每个函数也可以被多个 Action 调用，为多个视图服务。通过 Action 的处理，能实现函数和视图的灵活复用。

图 3.3 所示的是新 MVC 4 工程的目录结构。控制器层文件一般放置在 Controllers 文件夹

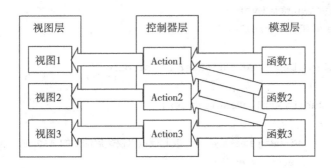

图 3.2 Action 映射示意图

下,一个控制器层文件对应一个控制器,以"控制器名称 + Controller"命名,图 3.3 中有 Home 和 Account 两个控制器。打开 HomeController.cs,其中的默认代码如下面的代码段所示。一个控制器就是一个特殊的类,在控制器顶部的代码是对命名空间的引用;控制器继承自 Controller 类;控制器中返回值是 ActionResult 的函数,一般都是 Action。当然,由于控制器是一个类,也可以在其中编写不是 Action 的函数。

图 3.3 控制器层文件示意图

HomeController.cs 的默认代码如下。

```
1    using System;
2    using System.Collections.Generic;
3    using System.Linq;
4    using System.Web;
```

```
5       using System.Web.Mvc;
6       namespace S3.Controllers
7       {
8           public class HomeController : Controller
9           {
10              public ActionResult Index()
11              {
12                  ViewBag.Message = "修改此模板以快速启动你的 ASP.NET MVC 应用程序。";
13                  return View();
14              }
15              public ActionResult About()
16              {
17                  ViewBag.Message = "你的应用程序说明页。";
18                  return View();
19              }
20              public ActionResult Contact()
21              {
22                  ViewBag.Message = "你的联系方式页。";
23                  return View();
24              }
25          }
26      }
```

注意1：在控制器中，若函数变成 Action，则必须满足下列条件：①函数必须是 public 的；②函数不能是静态的；③函数不能被重载；④函数不能是构造器（getter 或 setter）；⑤函数不能有开放式泛型类型（如 <T>）；⑥函数不能是基类的方法；⑦函数不能含有 ref、out。

注意2：Action 的返回值类型一般是 ActionResult 类型。由于 ActionResult 较为复杂，因此将其单独放在第6章介绍。

3.2 控制器的创建

当项目需求较为简单时，可以通过单个控制器来完成；但当项目需求较为复杂时，一般通过多个控制器来拆分功能，以降低每个功能的复杂度。在 MVC 4 中，一般通过下面的步骤来创建控制器。

1）选中"Controllers"文件夹，单击右键，出现如图 3.4 所示的快捷菜单。
2）将鼠标滑动到"添加"，出现如图 3.5 所示的子菜单。
3）单击"控制器"，出现如图 3.6 所示的"添加控制器"对话框。
4）修改控制器的名称（图 3.6 中选中部分），单击"添加"按钮后，将出现下面所示的 Default1Controller 初始代码。

第 3 章 控 制 器　63

图 3.4　右键单击"Controllers"后出现的快捷菜单

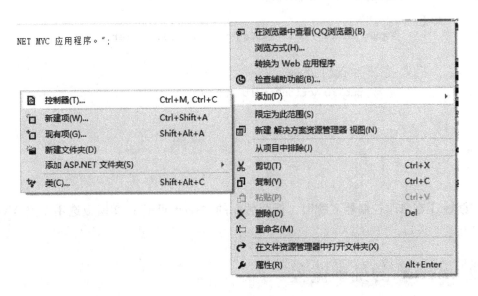

图 3.5　"添加"选项的子菜单

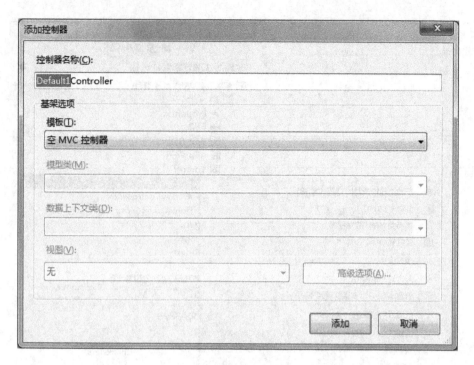

图3.6 "添加控制器"对话框

```
1    using System;
2    using System.Collections.Generic;
3    using System.Linq;
4    using System.Web;
5    using System.Web.Mvc;
6    namespace S3.Controllers
7    {
8        public class Default1Controller : Controller
9        {
10           // GET: /Default1/
11           public ActionResult Index()
12           {
13               return View();
14           }
15       }
16   }
```

建议：①在修改控制器名称时，保留后面的"Controller"；②模板选用"空 MVC 控制器"。

3.3 Action 的处理流程

视图层和模型层之间的交互是通过控制器层中的 Action 来完成的。一般每个 Action 在处理交互时都会经历 5 个处理流程，即参数获取、参数预处理、与模型层交互、结果预处理和视图返回。

当 Action 所要处理的功能较复杂时，参数预处理、与模型层交互、结果预处理可能会多次进行（如下文中的例 3.2 等）。而当出现下列情况时，可以省略相关的流程处理。

1) 无参数输入时，可以省略参数获取、参数预处理过程。
2) 输入到 Action 的参数和模型层相关函数的参数一致时，可以省略参数预处理过程。
3) 不需要和模型层交互时，可以省略与模型层交互过程。
4) 模型层返回的结果可以直接用于视图时，可以省略结果预处理过程。

本节将通过一个简单的例子，提醒读者每个处理流程需要注意的地方。

【例 3.1】通过学号查询学生姓名。

学生（Student）表的结构与例 2.1 一致。假设模型层中已存在数据库连接文件和有关函数（StudentRep.cs 中的 GetStudentByKey 函数）。本例尝试通过地址栏查询指定学号的男同学的姓名，如学号为 95001 的男生的姓名。

StudentRep.cs 代码如下。

```
1       using System;
2       using System.Collections.Generic;
3       using System.Linq;
4       using System.Web;
5       namespace S3.Models
6       {
7           public class StudentRep
8           {
9               StudentEntities db = new StudentEntities();    //数据库连接
10              public Student GetStudentByKey(string Sno,string Ssex)
11              {
12                  return db.Students.FirstOrDefault(s => s.Sno == Sno && s.Ssex == Ssex);
13              }
14          }
15      }
```

3.3.1 参数获取

在视图层和 Action 交互时，Action 一般通过两种方式获取有关数据（即参数）：一种是通过路由机制（将在第 4 章介绍）传递，有关参数将直接映射到 Action 的输入参数；另一种是通过表单传递（由于涉及视图层的设计，因此这种方式将在第 5 章介绍）。

在例 3.1 中，使用路由机制来实现参数的传递。因为需要传递的参数只有一个，所以可以默认路由来传递参数，故 Action 的初始构造如下列代码所示。在函数 SearchStudent 中，id 即为输出参数学号。

```
1       public ActionResult SearchStudent(string id)
2       {
3       }
```

3.3.2 参数预处理

参数的预处理主要包括两个方面：一个方面是将 Action 获取的参数转换成模型层函数能调用的参数，如字符串的拆分等；另一个方面是提供一些由 Action 中直接生成的参数的默认值，如例 3.1 中的查询的默认条件——性别为"男"。

在例 3.1 中，因为模型层的函数需要两个参数（即学号和性别），而 Action 只获取了学号这一个参数，所以性别需要设置默认值（即"男"），有关代码如下：

```
string sex = "男";
```

3.3.3 与模型层的交互

在 MVC 4 中，模型层中的文件一般是一个类，而一个业务逻辑就是类中的一个函数。因此，在控制器层与模型层交互时，实际上就是调用相关类中的函数。控制器层向模型层传递的数据就是函数的输入参数；而函数的返回值就是模型层向控制器层传递的参数。

调用模型层的函数一般需要 3 步：①在命名空间中调用相应的命名空间；②类的实体化；③函数的调用。

在例 3.1 中，首先要在顶部的命名空间编写如下代码。

```
using S3.Models;
```

然后，在控制器中定义如下全局变量（即类的实例化，方便在不同 Action 调用）。

```
StudentRep SR = new StudentRep();
```

最后，在相应的 Action 中使用已处理的参数调用该函数，如下列代码所示。

```
Student Stu = SR.GetStudentByKey(id,sex);
```

3.3.4 结果预处理

和参数预处理相似，结果预处理也主要包括两个方面：一个方面是将模型层返回的数据加工成视图层能使用的数据，如数据类型的转换、特定数据结构的组装等；另一个方面是将一些 Action 中直接生成的默认值提供给视图层，如视图层中一些控件的默认值。

在例 3.1 中，因为模型层的函数的返回值是 Student 类型，而功能要求的是姓名，所以需要提取其中的 Sname 字段。由于指定学号可能是女同学，会导致返回值不存在，因此需要进行异常处理。最终，结果预处理如下列代码所示。

```
1    string name = "";
2    try
3    {
4        name = Stu.Sname;
5    }
6    catch
7    {
8        name = "您查询的学生不存在!";
9    }
```

3.3.5 视图返回

Action 处理的结果是返回一个类型为 ActionResult 的视图，同时将有关数据传递给相应的视图。由于设计数据传递方式时涉及视图层的有关内容，因此将在第 5 章中详细介绍。ActionResult 较为复杂，将在第 6 章中单独介绍。

由于还未介绍视图层，因此例 3.1 中的返回值设定为文本类型，即 ContentResult。相应的处理过程如下列代码所示。

```
1    ActionResult result = Content(name);
2    return result;
```

3.3.6 实例分析

将上文中的有关代码合并，最终实现例 3.1 要求的功能的代码如下列代码（HomeController.cs）所示。其中，Index、About 和 Contact 这 3 个 Action 是系统默认生成的，SearchStudent 这个 Action 是例 3.1 所要实现的功能。

```
1    using System;
2    using System.Collections.Generic;
3    using System.Linq;
4    using System.Web;
5    using System.Web.Mvc;
6    using S3.Models;
7    namespace S3.Controllers
8    {
9        public class HomeController : Controller
10       {
11           StudentRep SR = new StudentRep();
12           public ActionResult Index()
13           {
14               ViewBag.Message = "修改此模板以快速启动你的 ASP.NET MVC 应用程序。";
15               return View();
16           }
17           public ActionResult About()
18           {
19               ViewBag.Message = "你的应用程序说明页。";
20               return View();
21           }
22           public ActionResult Contact()
23           {
24               ViewBag.Message = "你的联系方式页。";
25               return View();
26           }
27           public ActionResult SearchStudent(string id)
```

```
28                    {
29                        string sex = "男";
30                        Student Stu = SR.GetStudentByKey(id,sex);
31                        string name = "";
32                        try
33                        {
34                            name = Stu.Sname;
35                        }
36                        catch
37                        {
38                            name = "您查询的学生不存在!";
39                        }
40                        ActionResult result = Content(name);
41                        return result;
42                    }
43            }
44       }
```

按<F5>键运行程序后,在地址栏输入 http://localhost:端口号/Home/SearchStudent/学号,来查询相应学生的姓名。其中端口号是程序运行后自动配置的,学号是查询条件。假设端口号是16924,要查询的学号是95001,则查询结果如图3.7所示。端口号不变,若要查询的学号是95002,由于该学生是女生,则查询结果如图3.8所示。

图 3.7　学号为 95001 的查询结果

图 3.8　学号为 95002 的查询结果

3.4　典型的处理模式

在控制器中,Action 在处理视图层和模型层之间的交互时有 4 种典型的处理模式:①单个视图调用单个函数;②单个视图调用多个函数;③多个视图调用单个函数;④多个视图调用多个函数。图 3.2 示意了单个视图调用多个函数和多个视图调用单个函数的处理方式。

由于例 3.1 就是单个视图调用单个函数的处理方式,因此本节不再对这种处理模式进行说明。本节将通过例子来分别说明其余模式的处理方式。

3.4.1 单个视图调用多个函数

【例 3.2】 成绩查询 1。

该例数据库的结构与例 2.1 一致。假设模型层中已存在数据库连接文件。本例尝试通过地址栏查询指定学生的指定课程的成绩，如李名的数据库课程的成绩。

下面的 S3_2Rep.cs 代码展示了实现该功能所需要的模型层函数。由于选课（SC）表中只包含学号 Sno 和课程号 Cno，因此首先要获取这些信息。在学生（Student）表中，可以通过学生姓名找到该学生（为了简化示例，暂时不考虑重名，课程名称亦同），即 GetStudent 函数。之后，通过 Action 的处理可以得到学号 Sno。同理，在课程（Course）表中，可以通过课程名称得到该课程，即 GetCourse 函数。之后，通过 Action 的处理可以得到课程号 Cno。然后，通过 Sno 和 Cno 可以在选课（SC）表中找到相应的选课记录，即 GetSC 函数。最后，通过 Action 的处理可以得到相应的成绩。

S3_2Rep.cs 代码如下。

```
1     using System;
2     using System.Collections.Generic;
3     using System.Linq;
4     using System.Web;
5     namespace S3.Models
6     {
7         public class S3_2Rep
8         {
9             StudentEntities db = new StudentEntities();
10            public Student GetStudent(string name)
11            {
12                return db.Students.Where(s => s.Sname == name).FirstOrDefault();
13            }
14            public Course GetCourse(string name)
15            {
16                return db.Courses.Where(s => s.Cname == name).FirstOrDefault();
17            }
18            public SC GetSC(string sno, string cno)
19            {
20                return db.SCs.Where(s => s.Sno == sno && s.Cno == cno).FirstOrDefault();
21            }
22        }
23    }
```

下面列出的 HomeController.cs 中的关键代码展示了控制器中相应 Action 的处理过程。首先，根据学生姓名获取学号 Sno。接着，根据课程名称获取课程号 Cno。然后，根据学号 Sno 和课程号 Cno 获取选课记录。最后，根据选课记录获取成绩，并返回给视图。

本例中，在 Action 的处理过程中共调用 3 次模型层的函数。同时，为了控制异常，使用了多次 try...catch 结构进行异常处理。

HomeController.cs 中的关键代码如下。

```
1    S3_2Rep S32R = new S3_2Rep();
2    public ActionResult SearchGrade(string Sname, string Cname)
3    {
4        //获取 Sno
5        Student stu = S32R.GetStudent(Sname);
6        string Sno = "";
7        try
8        {
9            Sno = stu.Sno;
10       }
11       catch
12       {
13           return Content("您查询的学生不存在!");
14       }
15       //获取 Cno
16       Course cou = S32R.GetCourse(Cname);
17       string Cno = "";
18       try
19       {
20           Cno = cou.Cno;
21       }
22       catch
23       {
24           return Content("您查询的课程不存在!");
25       }
26       //获取成绩
27       SC sc = S32R.GetSC(Sno, Cno);
28       decimal grade = 0;
29       try
30       {
31           grade = sc.Grade;
32       }
33       catch
34       {
35           return Content("您查询成绩不存在!");
36       }
37       return Content(Sname + "的 " + Cname + " 的成绩为:" + grade);
38   }
```

按 <F5> 键运行程序后，通过在地址栏中输入 http://localhost:端口号/Home/SearchGrade?Sname=学生姓名&Cname=课程名称，来查询指定学生的指定课程的成绩。其中端口号是程序运行后自动配置的,学生姓名和课程名称是查询条件。假设端口号是 16924,要查询的学生姓名是李名,课程名称是数据库,则查询结果如图 3.9 所示。若查询的学生不存在、课程不存在或选课关系不存在,则会分别得到如图 3.10~图 3.12 所示的提示。

图 3.9 正常查询

图 3.10 学生不存在

图 3.11 课程不存在

图 3.12 选课关系不存在

注意: 由于示范的需要,因此本例进行了较为复杂的处理。实际上,可以在数据库中建立一个学生(Student)表、课程(Course)表和选课(SC)表关联的视图,然后使用单个视图调用单个函数的处理模式来处理。

3.4.2 多个视图调用单个函数

【例 3.3】学生信息查询。

学生(Student)表的结构与例 2.1 一致。假设模型层中已存在数据库连接文件。本例尝试通过地址栏分别查询指定学号的学生的姓名、性别、年龄、所在系所。

由于这些功能具有部分的共同特性(即查询指定学号的学生),因此模型层只需要一个函数,用于根据学号 Sno 获取学生记录,如 S3_3Rep.cs 中的 GetStudent 函数所示。

S3_3Rep.cs 代码如下。

```
1    using System;
2    using System.Collections.Generic;
3    using System.Linq;
```

```
4    using System.Web;
5    namespace S3.Models
6    {
7        public class S3_3Rep
8        {
9            StudentEntities db = new StudentEntities();
10           public Student GetStudent(string Sno)
11           {
12               return db.Students.FirstOrDefault(s => s.Sno == Sno);
13           }
14       }
15   }
```

在控制器层,分别编写4个不同的Action来完成相应的功能,每个Action分别用于提取指定的字段。在下列HomeController.cs的关键代码中,SearchName函数用于查询学生的姓名,SearchSex函数用于查询学生的性别,SearchAge函数用于查询学生的年龄,SearchDept函数用于查询学生的所在系所。

本例中,4个不同的视图均调用了一个函数。

HomeController.cs中的关键代码如下。

```
1    S3_3Rep S33R = new S3_3Rep();
2    public ActionResult SearchName(string id)
3    {
4        Student Stu = S33R.GetStudent(id);
5        try
6        {
7            return Content("学号为 " + id + " 的学生的姓名为:" + Stu.Sname);
8        }
9        catch
10       {
11           return Content("学号为 " + id + " 的学生不存在!");
12       }
13   }
14   public ActionResult SearchSex(string id)
15   {
16       Student Stu = S33R.GetStudent(id);
17       try
18       {
19           return Content("学号为 " + id + " 的学生的性别为:" + Stu.Ssex);
20       }
21       catch
22       {
23           return Content("学号为 " + id + " 的学生不存在!");
24       }
```

```
25      }
26      public ActionResult SearchAge(string id)
27      {
28          Student Stu = S33R.GetStudent(id);
29          try
30          {
31              return Content("学号为 " + id + " 的学生的年龄为:" + Stu.Sage);
32          }
33          catch
34          {
35              return Content("学号为 " + id + " 的学生不存在!");
36          }
37      }
38      public ActionResult SearchDept(string id)
39      {
40          Student Stu = S33R.GetStudent(id);
41          try
42          {
43              return Content("学号为 " + id + " 的学生的所在系别为:" + Stu.Sdept);
44          }
45          catch
46          {
47              return Content("学号为 " + id + " 的学生不存在!");
48          }
49      }
```

按<F5>键运行程序后,通过在地址栏中输入 http://localhost:端口号/Home/SearchName/学号,来查询相应学生的姓名。其中端口号是程序运行后自动配置的,学号是查询条件。假设端口号是 16924,要查询的学号是 95001,则查询结果如图 3.13 所示。同理,将 SearchName 分别换成 SearchSex、SearchAge、SearchDept 后,可以分别得到性别、年龄、所在系别的查询结果,分别如图 3.14~图 3.16 所示。

图 3.13 姓名查询结果

图 3.14 性别查询结果

图 3.15　年龄查询结果

图 3.16　所在系别查询结果

注意： 由于示范的需要，因此本例通过多个视图展示了同一个对象的不同属性。在实际的应用中，同一个对象的不同属性往往显示在一个视图中。

3.4.3　多个视图调用多个函数

【例 3.4】成绩查询 2。

该例数据库的结构与例 2.1 一致。假设模型层中已存在数据库连接文件。本例尝试通过地址栏分别查询：①指定学号的学生及其选修的全部课程和成绩；②指定课程号的课程名称以及选修该课的全部学生的姓名和成绩。

下列 S3_4Rep.cs 中的代码展示了实现该功能所需要的模型层函数。GetStudent 函数根据学号 Sno 获取学生记录；GetCourse 函数根据课程号 Cno 获取课程记录。由于两个功能中都涉及学生姓名和课程名称的查询，因此这两个函数会被多次调用。GetSCBySno 函数和 GetSCByCno 函数分别根据学号 Sno 和课程号 Cno 获取选课记录。由于这两个功能中查询的对象不一样，因此这两个函数会被分别调用。

S3_4Rep.cs 代码如下。

```
1    using System;
2    using System.Collections.Generic;
3    using System.Linq;
4    using System.Web;
5    namespace S3.Models
6    {
7        public class S3_4Rep
8        {
9            StudentEntities db = new StudentEntities();
10           public Student GetStudent(string Sno)
11           {
12               return db.Students.Where(s => s.Sno == Sno).FirstOrDefault();
13           }
14           public Course GetCourse(string Cno)
15           {
16               return db.Courses.Where(s => s.Cno == Cno).FirstOrDefault();
```

```
17          }
18          public IEnumerable<SC> GetSCBySno(string Sno)
19          {
20              return db.SCs.Where(s => s.Sno == Sno);
21          }
22          public IEnumerable<SC> GetSCByCno(string Cno)
23          {
24              return db.SCs.Where(s => s.Cno == Cno);
25          }
26      }
27  }
```

对于查询指定学号的学生及其选修的全部课程和成绩来说，可以分解为 3 个子步骤：①根据学号 Sno 查询学生姓名，即通过调用 GetStudent 函数获取；②通过学号 Sno 查询选课情况及相应的成绩，即通过调用 GetSCBySno 函数获取；③根据选课记录中的课程号 Cno 查询课程名称，即通过调用 GetCourse 函数获取。查询指定课程号的课程名称以及选修该课的全部学生的姓名和成绩的处理过程与之类似。下列 HomeController.cs 所示代码中的 SearchGradeBySno 和 SearchGradeByCno 这两个 Action 是相应功能的实现。

本例中，两个不同的 Action 调用了两个相同函数。

HomeController.cs 中的关键代码如下。

```
1   S3_4Rep S34R = new S3_4Rep();
2   public ActionResult SearchGradeBySno(string id)
3   {
4       //获取学生姓名
5       Student Stu = S34R.GetStudent(id);
6       string Sname = "";
7       try
8       {
9           Sname = Stu.Sname;
10      }
11      catch
12      {
13          return Content("学号为 " + id + " 的学生不存在!");
14      }
15      //获取选课记录(成绩)
16      IEnumerable<SC> SCs = S34R.GetSCBySno(id);
17      //获取课程名称
18      string result = "学号为 " + id + " 的学生的姓名为" + Sname + "。<br/>选修的
19      课程如下:<br/>";        //处理结果
20      if (SCs.Count() == 0)
21          result = "学号为 " + id + " 的学生未选修任何课程!";
22      foreach (SC sc in SCs)
23      {
```

```
24          Course Cou = S34R.GetCourse(sc.Cno);
25          result = result + Cou.Cname + ":" + sc.Grade + "<br/>";
26      }
27      return Content(result);
28  }
29  public ActionResult SearchGradeByCno(string id)
30  {
31      //获取课程名称
32      Course Cou = S34R.GetCourse(id);
33      string Cname = "";
34      try
35      {
36          Cname = Cou.Cname;
37      }
38      catch
39      {
40          return Content("课程号为 " + id + " 的课程不存在!");
41      }
42      //获取选课记录(成绩)
43      IEnumerable<SC> SCs = S34R.GetSCByCno(id);
44      //获取学生姓名
45      string result = "课程号为 " + id + " 的课程名称为" + Cname + "。<br/>选修
46  该课程的学生如下:<br/>";          //处理结果
47      if (SCs.Count() == 0)
48          result = "课程号为 " + id + " 的课程无任何学生选修!";
49      foreach (SC sc in SCs)
50      {
51          Student Stu = S34R.GetStudent(sc.Sno);
52          result = result + Stu.Sname + ":" + sc.Grade + "<br/>";
53      }
54      return Content(result);
55  }
```

按<F5>键运行程序后,通过在地址栏中输入 http://localhost:端口号/Home/SearchGradeBySno/学号,来查询相应学生的有关信息。其中端口号是程序运行后自动配置的,学号是查询条件。假设端口号是16924,要查询的学号是95001,则查询结果如图3.17

图3.17 学生信息查询结果

所示。同理，将 SearchGradeBySno 换成 SearchGradeByCno，将学号换成课程号 1 后，可以得到相应课程的有关信息，如图 3.18 所示。

图 3.18 课程信息查询结果

注意：由于示范的需要，因此本例进行了较为复杂的处理。实际上，同例 3.2 相似，可以在数据库中建立一个学生（Student）表、课程（Course）表和选课（SC）表关联的视图，然后使用单个视图调用单个函数的处理模式处理。

3.5 Action 的常见标签

在实现一些特定的功能时，往往要给 Action 函数添加一些标签，如 [HttpPost] 等。这些标签往往出现在 Action 函数的上一行，如下列代码所示。

```
1       [HttpPost]
2       public ActionResult Index(int id)
3       {
4           return View();
5       }
```

Action 的标签有很多种，其中一些常见标签见表 3.1。

表 3.1　Action 的常见标签

标签名称	使用方式	作用
ActionName	[ActionName("新名字")]	新名字将会取代原来的 Action 名字（Anytao, 2009）
NonAction	[NonAction]	对应的 Action 不会被访问到
HttpGet	[HttpGet]	对应的 Action 只响应 HTTP-Get 请求
HttpPost	[HttpPost]	对应的 Action 只响应 HTTP-Post 请求
Authorize	[Authorize]	只有登录后才能访问对应的 Action（Johnny Yan, 2012）
Authorize	[Authorize(Roles = "角色名")]	只有指定角色登录后才能访问对应的 Action
Authorize	[Authorize(Users = "用户名")]	只有指定用户登录后才能访问对应的 Action
ChildActionOnly	[ChildActionOnly]	对应的 Action 只能被子请求访问到
RequireHttps	[RequireHttps]	对应的 Action 只能被 HTTPS 方式访问，若通过 HTTP 方式访问，将跳转到 HTTPS 方式
ValidateRequest	[ValidateRequest(true/false)]	默认情况下为 true，表示对传入的参数进行验证，预防 XSS 攻击（苏飞, 2009）
ValidateAntiForgeryToken	[ValidateAntiForgeryToken]	预防 CSRF 攻击（2011）
OutputCache	[OutputCache(参数 = 参数值)]	对对应的 Action 的结果进行缓存（sslyc8991, 2013）

下面将介绍其中比较常用的几种标签。

3.5.1 NonAction

NonAction 的作用是阻止相应的 Action 被访问到。下列代码是正常的 Action，在程序运行后，访问地址/Home/ NonActionTest1，可以得到如图 3.19 所示的正常的访问结果。当加上［NonAction］标签后（见下列带［NonAction］标签的 Action 所示的代码），访问地址/Home/ NonActionTest2，会得到如图 3.20 所示的访问异常错误，即找不到该 Action。

正常的 Action 的代码如下。

```
1       public ActionResult NonActionTest1()
2       {
3           return Content("如果出现这段话,表示该 Action 能被访问到!");
4       }
```

带［NonAction］标签的 Action 的代码如下。

```
1       [NonAction]
2       public ActionResult NonActionTest2()
3       {
4           return Content("如果出现这段话,表示该 Action 能被访问到!");
5       }
```

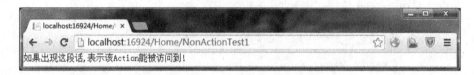

图 3.19 正常的访问结果

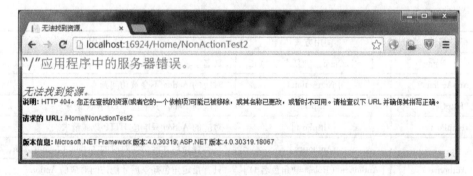

图 3.20 访问异常错误

NonAction 相当于把 Action 的访问权限从 public 变成了 private。在下列代码中，Action 没有添加［NonAction］标签，但将它的访问权限改成了 private，因此，在访问地址/Home/ NonActionTest3 时，仍然会得到如图 3.20 所示的访问异常错误。

```
1       private ActionResult NonActionTest3()
2       {
3           return Content("如果出现这段话,表示该 Action 能被访问到!");
4       }
```

NonAction 的主要用途是屏蔽一些当前版本不希望被用户使用的 Action，其优势在于不必注释整个 Action，便于查看和修改该 Action。

3.5.2　HttpGet 和 HttpPost

一般情况下，［HttpGet］标签和［HttpPost］标签会配对使用，两者所对应的 Action 的名字相同，如下面列出的［HttpGet］标签和［HttpPost］标签配对使用的代码所示。这两个标签主要用于表单的处理：使用了［HttpGet］标签的 Action 主要用于处理表单页面的生成；使用了［HttpPost］标签的 Action 主要用于接收相应的表单数据，并对其进行处理。处理完成后，一般跳转到另外的 Action。

［HttpGet］标签和［HttpPost］标签的配对使用代码与 FormTest.cshtml 中的代码是对表单处理的一个简单示例，即在用户输入姓名并提交后，显示用户输入的姓名。前者中的第一个 Action 用于生成如图 3.21 所示的页面，由于功能较为简单，因此该 Action 无其他代码；而有关的页面布局在后者所示的代码（视图层文件）中。前者中的第二个 Action 用于处理表单数据，即提取用户输入的姓名。为了简化代码，本例没有使用单独的 Action 进行输出处理，而在第二个 Action 中直接输出了。（有关视图层的内容，以及与模型层之间的交互，将在第 5 章介绍。）

［HttpGet］标签和［HttpPost］标签的配对使用代码如下。

```
1       [HttpGet]
2       public ActionResult FormTest()
3       {
4           return View();
5       }
6       [HttpPost]
7       public ActionResult FormTest(FormCollection collection)
8       {
9           string name = collection["name"];
10          return Content("您的名字是:" + name);
11      }
```

FormTest.cshtml 中的代码如下。

```
1       @{
2           Layout = null;
3       }
4       <!DOCTYPE html>
5       <html>
6       <head>
7           <meta name="viewport" content="width=device-width" />
8           <title>FormTest</title>
9       </head>
10      <body>
11          <div>
12              <form action="/Home/FormTest" method="post">
```

```
13          请输入您的姓名:<input name = "name">
14          <input type = "submit" value = "提交"/>
15      </form>
16    </div>
17  </body>
18 </html>
```

程序运行后，通过地址栏访问/Home/FormTest，将得到如图3.21所示的页面。在输入姓名（见图3.21中的"张三"），并单击"提交"按钮后，将得到如图3.22所示的结果。

图3.21 HttpGet 返回的页面

图3.22 HttpPost 返回的页面

注意：根据实际需求的不同，[HttpGet] 标签和 [HttpPost] 标签可以单独分别使用，两者所对应的 Action 的名字可以不相同。

3.5.3 ChildActionOnly

ChildActionOnly 的作用是使对应的 Action 只能被子请求访问到，即不能直接访问。下面列出的 [ChildActionOnly] 标签的使用代码是对 [ChildActionOnly] 标签使用的简单示例。程序运行后，通过地址栏访问/Home/ChildTest2，将得到如图3.23所示的访问异常。而 ChildTest.cshtml 中的代码是 ChildTest 的视图层代码（有关视图层的内容将在第5章介绍），其中通过@ Html.Action("ChildTest2")实现了对 ChildTest2 的子请求访问。通过地址栏访问/Home/ChildTest，将得到如图3.24所示的正常结果。本例中是希望 ChildTest2 仅作为分部页被访问，而不是作为视图页被访问，因此，为 ChildTest2 添加了 [ChildActionOnly] 标签。

[ChildActionOnly] 标签的使用代码如下。

```
1      public ActionResult ChildTest()
2      {
3          return View();
4      }
5      [ChildActionOnly]
```

```
6        public ActionResult ChildTest2()
7        {
8            return Content("只能通过子请求访问到本内容!");
9        }
```

ChildTest.cshtml 中的代码如下。

```
1        @{
2            Layout = null;
3        }
4        <!DOCTYPE html>
5        <html>
6        <head>
7            <meta name="viewport" content="width=device-width" />
8            <title>ChildTest</title>
9        </head>
10       <body>
11           <div>
12               @Html.Action("ChildTest2")
13           </div>
14       </body>
15       </html>
```

图 3.23　直接访问异常

图 3.24　子请求访问正常

※习　题

1. 新建一个名为"Mine"的控制器。
2. 建立一个学生（Student）表、课程（Course）表和选课（SC）表关联的视图，使用单个视图调用单个函数的处理模式实现例 3.2 的功能。
3. 建立一个学生（Student）表、课程（Course）表和选课（SC）表关联的视图，使用单个视图调用单个函数的处理模式实现例 3.4 的功能。

※综合应用

本书将在第 2、3 和 5 章的综合应用中建立一个基本的拥有购物车功能的网站。其中，此处模块的重点主要在于 Model 的编写。

在第 2 章综合应用的基础上，添加控制层模块。控制层模块主要分为 BaseController.cs、CartController.cs、MemberController.cs 和 OrderController.cs 4 个类。

BaseController.cs 是基础类，用于存储购物车的列表。

```
1       protected MvcShoppingContext db = new MvcShoppingContext();
2       protected List<Cart> Carts
3       {
4           get
5           {
6               if (Session["Carts"] == null)
7               {
8                   Session["Carts"] = new List<Cart>();
9               }
10              return (Session["Carts"] as List<Cart>);
11          }
12          set { Session["Carts"] = value; }
13      }
14  }
```

CartController.cs 用于控制购物车中的内容，包括初始化购物车、向购物车中添加物品、将物品从购物车中移除、查看购物车中商品的数量等信息。

```
1       //加入产品项目到购物车,如果没有传入 Amount 参数,则预设购买数量为 1
2       [HttpPost] //因为知道要通过 AJAX 调用这个 Action,所以可以先标示 [HttpPost] 属性
3       public ActionResult AddToCart(int ProductId, int Amount = 1)
4       {
5           var product = db.Products.Find(ProductId);
6           //验证产品是否存在
7           if (product == null)
8               return HttpNotFound();
9           var existingCart = this.Carts.FirstOrDefault(p => p.Product.Id == ProductId);
```

```csharp
10              if (existingCart ! = null)
11              {
12                  existingCart.Amount + = 1;
13              }
14              else
15              {
16                  this.Carts.Add(new Cart() { Product = product, Amount = Amount });
17              }
18              return new HttpStatusCodeResult(System.Net.HttpStatusCode.Created);
19          }
20          ...
21          return RedirectToAction("Index", "Cart");
22      }
```

MemberController.cs 为会员注册的控制层，主要功能是实现用户的注册、登录和登出等操作。

```csharp
1   public class MemberController : BaseController
2   {
3       //会员注册页面
4       public ActionResult Register()
5       {
6           return View();
7       }
8       //密码散列所需的 Salt 随机值
9       private string pwSalt = "A1rySq1oPe2Mh784QQwG6jRAfkdPpDa90J0i";
10      //写入会员资料
11      [HttpPost]
12      public ActionResult Register ([Bind (Exclude = " RegisterOn, Auth
13  Code")] Member member)
14      {
15          //检查会员是否已存在
16          var chk_member = db.Members.Where(p = > p.Email == member.Email).Fir-
17  stOrDefault();
18          if (chk_member ! = null)
19          {
20              ModelState.AddModelError("Email", "您输入的 Email 已经有人注册
21  过了!");
22          }
23      private bool ValidateUser(string email, string password)
24      {
25          var hash_pw = FormsAuthentication.HashPasswordForStoringIn Con-
26  figFile(pwSalt + password, "SHA1");
27          var member = (from p in db.Members
28                        where p.Email == email && p.Password == hash_pw
```

```csharp
29                       select p).FirstOrDefault();
30             //如果member不为null,则代表会员的账号、密码输入正确
31             if (member != null) {
32                 if (member.AuthCode == null) {
33                     return true;
34                 } else {
35                     ModelState.AddModelError("", "您尚未通过会员验证,请收信并
36 点击会员验证链接!");
37                     return false;
38                 }
39             } else {
40                 ModelState.AddModelError("", "您输入的账号或密码错误");
41                 return false;
42             }
43         }
44         [HttpPost]
45         public ActionResult CheckDup(string Email)
46         {
47             var member = db.Members.Where(p => p.Email == Email)
48 .FirstOrDefault();
49             if (member != null)
50                 return Json(false);
51             else
52                 return Json(true);
53         }
54     }
```

OrderController.cs用于显示表单,并将资料存入数据库。

```csharp
1  [Authorize] //必须登录,会员才能使用订单结账功能
2  public class OrderController : BaseController
3  {
4      //显示完成订单的表单页面
5      public ActionResult Complete()
6      {
7          return View();
8      }
9      //将订单资料与购物车资料存入数据库
10     [HttpPost]
11     public ActionResult Complete(OrderHeader form)
12     {
13         var member = db.Members.Where(p => p.Email == User.Identity.Name).
14 FirstOrDefault();
15         if(member == null) return RedirectToAction("Index", "Home");
16         if (this.Carts.Count == 0) return RedirectToAction("Index", "Cart");
```

```
17                db.Orders.Add(oh);
18                db.SaveChanges();
19                //订单完成后必须清空现有购物车资料
20                this.Carts.Clear();
21                //订单完成后回到网站首页
22                return RedirectToAction("Index", "Home");
23            }
24        }
```

下面是完整对应的登录页面，可与注册页面一同测试代码执行情况。

```
1   @model MvcShopping.Models.MemberLoginViewModel
2   <h2>会员登录</h2>
3   @using (Html.BeginForm()) {
4       @Html.ValidationSummary(true)
5   <fieldset>
6   <legend>请输入您的账号、密码</legend>
7   <div class="editor-label">
8           @Html.LabelFor(model => model.email)
9   </div>
10  <div class="editor-field">
11          @*@Html.EditorFor(model => model.email)*@
12          @Html.TextBoxFor(model => model.email, new { data_val_email = "
13  请输入Email地址" })
14          @Html.ValidationMessageFor(model => model.email)
15  </div>
16  <div class="editor-label">
17          @Html.LabelFor(model => model.password)
18  </div>
19  <div class="editor-field">
20          @Html.EditorFor(model => model.password)
21          @Html.ValidationMessageFor(model => model.password)
22  </div>
23  <p>
24  <input type="submit" value="登录" />
25  </p>
26  </fieldset>
27  }
28  <div>
29      @Html.ActionLink("回首页", "Index", "Home")
30  </div>
31  @section Scripts {
32      @Scripts.Render("~/bundles/jqueryval")
33      @if (TempData["LastTempMessage"] != null)
34      {
```

```
35    <script>
36    alert('@HttpUtility.JavaScriptStringEncode(Convert.ToString(TempData["
37    LastTempMessage"]))');
38    </script>
39        }
40    }
```

第 4 章

路　由

本章主要介绍什么是路由，路由的解析、注册与管理以及 ASP.NET MVC 执行生命周期等重要内容，其中路由解析、注册和 ASP.NET MVC 执行生命周期尤为重要。要想掌握 ASP.NET MVC，本章内容必须认真学习。本章通过如何寻找函数、如何在 ASP.NET MVC 中寻找跳转网页等重要内容，加深对 ASP.NET MVC 执行时的先后顺序的理解，进而减少出错的机会。

4.1 路由的基础

4.1.1 网址路由的作用

在 ASP.NET MVC 框架下，网页寻址的方式是通过路由实现的。因此，网址路由（即 Routing）在 ASP.NET 中有两个主要用途：一个用途是它对比通过浏览器传入的 HTTP 请求，另一个用途是它将适当的网址返回给浏览器进行显示。

1. 对比 HTTP 请求

在客户端对 ASP.NET MVC 服务器发出 HTTP 请求时，是通过 Routing 找到适当的 HttpHandler 来处理页面，具体过程如图 4.1 所示。

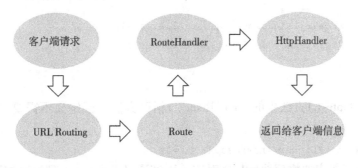

图 4.1　客户端对 ASP.NET MVC 服务器的请求流程

对 HttpHandler 的处理方法有两种，即 ASP.NET 网址方法处理和由 MvcHandler 来处理。如果由 MvcHandler 来处理，则会进入 ASP.NET MVC 执行生命周期，通过寻找相应的 Controller 与 Action 来对所发送的请求进行处理，并将处理后的信息返回给客户端。

2. 将适当的网址返回浏览器

网址路由在处理请求后还决定了 ASP. NET MVC 应该返回什么样的网址给客户端使用。一般情况下，页面地址跳转或者 View 中显示超链接时，都需要参考 Routing 的定义，以决定 ASP. NET MVC 应该输出什么样的网址。

4.1.2 默认的 Route Table

当创建一个新的 ASP. NET MVC 应用程序时，这个应用程序已经被配置用来使用 ASP. NETRouting。ASP. NET Routing 需要在以下两个地方进行设置。

第一，ASP. NET Routing 在用户的应用程序的 Web 配置文件（即 Web. config 文件）中是有效的。在配置文件中，有 4 个与 Routing 相关的代码片段：system. web. httpModules 代码段、system. web. httpHandlers 代码段、system. webserver. modules 代码段及 system. webserver. handlers 代码段。如果没有这些代码段，那么 Routing 将不再运行。

第二，Route Table 在应用程序的 App_Start/RouteConfig.cs 文件中创建。这个 RouteConfig.cs 文件是一个特殊的文件，它包含 ASP. NET 应用程序生命周期 events 的 eventhandlers，这个 Route Table 在应用程序的起始 event 中创建。

在 Global. asax 文件中已经定义了两个默认的网址路由，代码如下，下面将进行详细讲解。

```
1    public static void RegisterRoutes(RouteCollection routes)
2    {
3        routes.IgnoreRoute("{resource}.axd/{*pathInfo}");
4        routes.MapRoute(
5            "Default", //路由名称
6            "{controller}/{action}/{id}", //带有参数的 URL
7            new { controller = "Home", action = "Index", id = UrlParameter.Optional }
8        );
9    }
10   protected void Application_Start()
11   {
12       AreaRegistration.RegisterAllAreas();
13       RegisterRoutes(RouteTable.Routes);
14   }
```

其中：

1) RegisterRoutes()方法中的 IgnoreRoute()辅助方法用于定义不需要通过 Routing 处理的网址，例如：

Http://localhost/Dinner/index.aspx?id=12

本地址由于有相应的物理地址与之对应，不需通过 Routing 处理而直接可以得到，因此执行 RegisterRoutes()方法。RegisterRoutes()是 ASP. NET MVC 4(System. Web. Mvc)的一部分。

2) {resource} 代表一个名为"resource"的 RouteValue 表达式，它代表一个 PlaceHolder 类的变量空间，用于放置一个用不到的变量。因此，这里可以是任意名字。

3）｛*pathInfo｝代表一个名为"pathInfo"的 RouteValue 表达式。"pathInfo"前的"*"表示取得所有的值。"pathInfo"所表示的 RouteValue 表达式的值是完全路径信息中除去｛resource｝中的部分后剩余的部分网址。例如，网址为"/Dinner.aspx/id/book"，这里"resource"代表的是"Dinner.aspx"，｛*pathInfo｝的值为"id/book"。如果只有｛pathInfo｝，没有"*"，则其值为"id"。这里 pathInfo 只是一个符号，可以换成其他符号。

4）MapRoute()方法是用来定义 Routing 规则的辅助方法。MapRoute()是 ASP.NET MVC（System.Web.Mvc）的一部分。其定义了路由的形式、默认值、规则和返回形式等。

5）"Default"定义 Route 的名称，即这里的 Route 的名称为"Default"。

6）"｛controller｝/｛action｝/｛id｝"定义 Route 的寻址格式和每个网址段落的 RouteValue 表达式名称。该网址不能用"/"开头。

7）new 定义了各个路由的默认值，当网址路由对比不到 HTTP 请求时，就会改以默认值替代。

在客户端发送一个请求后，所有 ASP.NET Web 应用程序执行的入口都为 HttpApplication 的 Application_Start()事件，所有的 Routing 都会在此定义。其中，RouteTable.Routes 是一个公开的静态对象，用于存储所有 Routing 的规则集。

因此，当 ASP.NET 应用程序第一次启动时，即会调用 Application_Start()方法。此方法将调用 RegisterRoutes()方法创建路由表。默认路由表包含一个路由（名称为 Default）。Default 路由将 URL 的第一段映射到控制器名称，将第二段映射到控制器操作，将第三段映射到名称为 id 的参数。

【例4.1】假设将下面的 URL 输入到 Web 浏览器的地址栏：

`/Home/Index/3`

Default 路由将此 URL 映射为下列参数：

`controller = Home`
`action = Index`
`id = 3`

在请求 URL：/Home/Index/3 时，执行下面的代码：

`HomeController.Index(3)`

Default 路由包括三个参数的默认值。如果不提供控制器，则控制器参数默认为值 Home。如果不提供操作，则操作参数默认为值 Index。最后，如果不提供 id，则 id 参数默认为空字符串。

【例4.2】在默认生成的工程中，输入以下 URL，观察页面结构。

1）/Home/Index/3
2）/Home/Index/6
3）/Home/Index/name
4）/Home/Index/@@

4.2 路由解析

ASP.NET MVC 4 提供了强大的 URL 路由引擎，可以很好地控制 URL 映射到控制器。它允许我们完全定制 ASP.NET MVC 如何选择 controller 类、调用哪一个方法，以及从 URL 中自动解析变量值并作为参数传递给方法。

4.2.1 非 MVC 控制器类路由解析

对于在 ASP 技术或 JSP 技术中利用网址和"?"来传递参数的做法，是无法正常使用 MVC Router 技术解析的。这时需要一个非 MVC Router 类的网址的入口进而接收这类路由，这就是 IgnoreRoute 方法。该方法格式如下：

routes.IgnoreRoute("{resource}.axd/{*pathInfo}");

该方法可以解析 MVC Router 无法解析的路由地址。例如，http://localhost/index/login.aspx?id='1'，这样的地址由于有问号的存在，因此是无法通过 routes.MapRoute 路由的，但是可以通过 IgnoreRoute 路由，进而完成 MVC 中各种路由形式共同创建的功能。

注意： 1) http://localhost/index/login.aspx?id='1'通过 IgnoreRoute 路由是直接访问 login.aspx 页面，而不会访问控制层。

2) 当一个页面有真实地址时，也可以直接通过 IgnoreRoute 访问该页面。

4.2.2 带单个参数的 MVC 路由地址的解析

该路由形式在通过 MVC Router 进行路由的同时，传递一个参数。这种方法将根据用户的函数名称和参数到控制层中相应的方法中执行相应的代码，并返回与方法有相同名称的页面。

例如，MVC 默认路由：

```
1    routes.MapRoute(
2            "Default", //路由名称
3            "{controller}/{action}/{id}", // 带有参数的 URL
4            new { controller = "Home", action = "Index", id =
5    UrlParameter.Optional } //参数默认值
6            );
```

这就是单个参数的 MVC Router 形式。其中 controller 是控制器名称，action 是方法名，id 为要传递的参数。

【例 4.3】表 4.1 展示了 URL 地址是如何使用默认的 MVC Router 规则进行映射的。

表 4.1 默认的 MVC Router 网址映射

URL	Controller Class	Action Method	Parameter Passed
/Dinners/Details/2	DinnersController	Details(id)	id = 2
/Dinners/Edit/5	DinnersController	Edit(id)	id = 5
/Dinners/Create	DinnersController	Create()	
/Dinners	DinnersController	Index()	
/Home	HomeController	Index()	
/	HomeController	Index()	

注意：1）这里的 id 可以是任意类型（整型、字符串、浮点数等），具体类型由在 action 中的参数类型而定。

2）action 中的参数名必须是 id，否则无法找到相应的函数。

4.2.3 带多参数的 MVC 路由地址的解析

多参数即指两个以上的参数。在 MVC 中，这样传递参数的方法有两种：

1）使用网址加问号的形式传递多参数。

例如：http：//localhost/index/login.aspx？id = '1'&password = '1'。

这种方式的传递在 4.2.1 节中已经介绍过。

2）使用多参数路由。

例如：

```
1    routes.MapRoute(
2                "Default", //路由名称
3                "{controller}/{action}/{id}/{password}", //带有参数的 URL
4                new { controller = "Home", action = "Index", id = "11",pass-
5    word = "22" } // 参数默认值
6                );
```

该路由是双参数路由，配置控制层相应的方法，此方法有两个参数，即 id 和 password，其中，new 中的是参数的默认值。

【例 4.4】 表 4.2 展示了 URL 地址是如何使用以上的 MVC Router 规则进行映射的。

表 4.2 自定义双参数的 MVC Router 网址映射

URL	Controller Class	Action Method	Parameter Passed
/Dinners/Details/2/1	DinnersController	Details（id, password）	id = 2，password = 1
/Dinners/Edit/5/5	DinnersController	Edit（id, password）	id = 5，password = 5
/Dinners/Create	DinnersController	Create（）	id = 11，password = 22

注意：1）无论有几个参数，都需要如"{controller}/{action}/{id}/{password}"的形式定义参数的名称和先后顺序。同时，action 中的参数的名称和先后顺序也要和这里的完全一样。

2）这里的每一个参数都可以是任意类型（整型、字符串、浮点数等），具体类型由在 action 中的参数类型而定。

3）MVC Router 可以有多个参数，但是各参数的名称不能相同。

4.3 路由注册

到目前为止，我们已经讲了什么是路由，以及浏览器是如何通过路由解析到我们想要的

页面的。现在，我们来学习如何自定义路由，即路由注册。

顾名思义，路由注册就是将我们想要的而目前系统没有的路由注册到我们的项目中，以达到可以使用这种形式的路由的目的。最常见的就是注册多参数的路由。下面以一个多参数路由注册的例子说明路由注册的方法。

【例 4.5】 所要传递的路径值为：NewExample/login/linder/12/女/第一中学。

1）打开 Global.asax.cs 文件，在其中添加如下代码。

```
1       routes.MapRoute(
2       "login",    //路由名称
3       "{controller}/{action}/{name}/{age}/{sex}/{school}",    //有 4 个参数的 URL
4       new { controller = "Home", action = "Index" } //参数默认值
5       );
```

注意：本段代码建立了一个新的路由，路由名称为"login"，其中带有 4 个参数，分别是 name、age、sex 和 school，并且这 4 个参数的位置可以是任意的。

2）建立一个名为 NewExample 的控制层，在其中建立一个名为 login 的方法。具体代码为：

```
1       public ActionResult login(string name, int age, string sex, string school)
2       {
3           ViewData["name"] = name;
4           ViewData["age"] = age;
5           ViewData["sex"] = sex;
6           ViewData["school"] = school;
7           return View();
8       }
```

注意：这段代码中的参数的个数要和路由中配置的参数个数完全相同，并且各个参数的先后顺序要和路由中的一致。另外，参数名也必须完全相同，不能有半点偏差。

3）建立相应的显示页面，显示传递过来的参数，主体代码如下。

```
1       姓名：<% :ViewData["name"].ToString() %>
2       <br /><hr /><br />
3       年龄：<% :ViewData["age"].ToString() %>
4       <br /><hr /><br />
5       性别：<% :ViewData["sex"].ToString() %>
6       <br /><hr /><br />
7       学校：<% :ViewData["school"].ToString() %>
```

4）在母版也建立相应的连接页，代码为：

```
<% : Html.ActionLink("例子", "login/linder/12/女/第一中学", "NewExample")% >
```

注意：在方法的位置可以直接跟上参数，路由后的地址即为：

NewExample/login/linder/12/女/第一中学

5）最后在网页上打开，显示结果如图 4.2 所示。

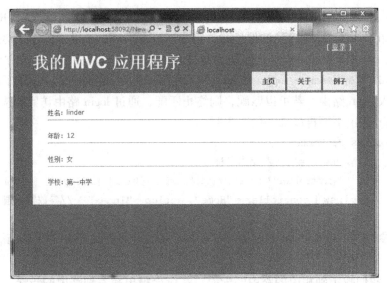

图 4.2 "NewExample/login/linder/12/女/第一中学"的页面效果

4.4 路由管理与匹配机制

路由是在 App_Start/RouteConfig.cs 文件中注册的。通过上文的介绍，我们知道了它的默认值、匹配了哪种形式的输入、返回到何处等。但是，当一个地址传递过来后，它是以什么样的次序匹配路由的呢？

实际上，Global.asax.cs 中的路由匹配是按照自上而下的先后顺序进行匹配的。当用户输入一段地址时，它将从上而下开始匹配，直到第一个可以匹配的路由为止。

注意：如果两个路由的形式相同（参数的个数等），那么它将只匹配上面的一个路由而不匹配下面的一个路由。

【例 4.6】有如下一段 Global.asax.cs 的代码：

```
1    public static void RegisterRoutes(RouteCollection routes)
2    {
3        routes.IgnoreRoute("{resource}.axd/{*pathInfo}");
4        routes.MapRoute(
5            "Default", //路由名称
6            "{controller}/{action}/{id}", //带有参数的 URL
7            new { controller = "Home", action = "Index", id = UrlParameter.Optional } //
8    参数默认值
9        );
10       routes.MapRoute(
11           "login",   //路由名称
12           "{controller}/{action}/{name}/{age}/{sex}/{school}",   //有4个参数的URL
13           new { controller = "Home", action = "Index" } //参数默认值
14       );
15   }
```

当输入一段 ASP.NET 代码或有物理地址的代码时,它将直接匹配 IgnoreRoute 路由。

当输入如 "Hello/login/2" 的地址时,首先匹配 IgnoreRoute 路由,若无法匹配,则匹配 Default 路由,若可以匹配,则终止匹配,通过 Default 路由访问相应的函数方法。

当输入如 "NewExample/login/linder/12/女/第一中学" 的地址时,若前两个地址均无法匹配,则进入 login 路由,若可以匹配,则终止匹配,通过 login 路由访问相应的函数方法。

当在 login 路由下面再添加一段路由时:

```
1    routes.MapRoute(
2        "Double",    //路由名称
3        "{controller}/{action}/{id}/{number}/{home}/{college}",    //有4个参数的URL
4        new { controller = "Home", action = "Index" } //参数默认值
5    );
```

由于其参数的个数和 login 的路由完全相同,因此在匹配之后用到 login 路由而达不到该路由。

注意:当我们找不到相应的路由匹配时,系统会报出找不到路由的错误。

4.5 MVC 执行的生命周期

ASP.NET MVC 执行的生命周期大致分为三个阶段,即:

1)网址路由比对阶段。
2)执行 Controller 的 Action 阶段。
3)执行 View 并返回结果页面。

每一个 MVC 网站都以这样的顺序执行每一个网址请求。下面我们将详细讲解每一个阶段。

4.5.1 网址路由比对阶段

一个请求通过 IIS 发送后,先要经过 UrlRoutingModule 处理与网址路由相关的运算。默认情况下,如果在网站的目录下有实体文件与之对应,则不经过 MVC 的 Routing 处理,而是直接交由 IIS 或 ASP.NET 进行处理。

但是,当所请求的文件不存在时,ASP.NET MVC 的 Routing 就会正式启动对比,并对比这一网址。例如,对于 "http://localhost/Member/Login.aspx" 这一网址,由于网站中有实体文件,因此会直接通过 IIS 跳转到 Login.aspx 文件执行,不会经过 MVC 路由;但是,当 Login.aspx 文件不存在时,将启动 MVC 路由,通过比对路由来比对相应的 action 函数,将执行 MemberController 下面的 Login 函数。

如果我们在 Global.aspx 文件的 Application_Start() 实现的最前面将 RouteTable.Routes.RouteExistingFiles 参数的值设置为 "true",那么网站将不会先判断是否有实体文件存在。相应代码如下:

```
1    protected void Application_Start()
2    {
3        RouteTable.Routes.RouteExistingFiles = true;
```

```
4                    AreaRegistration.RegisterAllAreas();
5                    RegisterRoutes(RouteTable.Routes);
6          }
```

设定完成后，该网站收到的所有 HTTP 请求都会使用 RegisterRoutes( ) 方法中定义的网址路由进行比对，若比对成功，则将使用 MVC 进行处理，若比对失败，则交还给 IIS 进行处理。

注意：当使用 RegisterRoutes( ) 中的 IgnoreRoute 进行处理时，实际上也是通过 IIS 对网址进行处理的，它不经过 MVC 路由。

4.5.2　执行 Controller 的 Action 阶段

当网址比对成功后，将进入这一阶段，代码将执行 MvcHander 类。MvcHander 会通过 ProcessRequestInit( ) 方法先用 ControllerBuilder 对象取得一个 IControllerFactory 接口，然后再通过该接口建立 IController 接口。最后返回 ProcessRequest( ) 方法，执行该 Controller 的 Execute( ) 方法。

注意：在 ASP.NET MVC 中，默认的 IControllerFactory 接口对象为 DefaultControllerFactory，但是我们可以通过自定义实现 IControllerFactory 接口的类获得 Controller 层导向。

另外，由于 MVC 默认的所有在"Controllers"目录下的 Controller 都会继承 System.Web.Mvc.Controller 类，而它又继承自 System.Web.Mvc.ControllerBase 类，而该类下有 Execute( ) 方法，因此执行过程最后会再一次调用 System.Web.Mvc.Controller 类的 ExecuteCore( ) 方法。当执行 System.Web.Mvc.ControllerBase 类的 ExecuteCore( ) 方法时，会通过路由值决定要执行 Controller 里的哪个 Action。

注意：如果找不到需要的 Action，那么将会执行 HandleUnknownAction( ) 方法。

在找到相应的 Action 并执行后，会返回相应的页面。Action 默认的响应类是 ActionResult 类。ActionResult 类是一个抽象类，将在第 5 章讲解。

4.5.3　执行 View 并返回结果页面

当从 Action 返回的 ActionResult 对象为 ViewResult 对象时，MVC 会进一步调用 IViewEngine 接口的对象实体的 FindView( ) 方法，以取得一个实现 IView 接口的对象实体，然后再调用 IView 对象实体的 Render( ) 方法，将相应内容返回客户端。默认的 Iview 接口的实体是 WebForm View 类。该类会建立 ViewPage 对象，而 ViewPage 类则继承自 System.Web.UI.Page 类。

注意：建立相应的 ViewPage 并不等于使用所有的 WebForm 功能。如果试图使用 MVC 不支持的 Page 功能，那么会导致异常状态。

4.6　总结

路由是 ASP.NET MVC 的一个核心步骤，没有路由，请求将不知道发送到什么位置。虽然在 MVC 下支持 ASP.NET 方式的网址路由和 MVC 形式的网址路由，但是为了更好地匹配 MVC 这一框架，建议尽量使用 MVC 路由。

※ 习 题

1. MVC 中路由的作用是什么？MVC 中的路由和直接应用路径的做法的区别在哪里？
2. 编写一个新的路由，可以传递 3 个参数，并使用页面实现传递的结果。

※ 综合应用

1. 综合运用路由协议的方法。在一个新的项目中，创建一个名为 MapInfoAction 的路由，对其进行重新复用，使其可以不带参数使用，带一个参数使用，带多个（2 个以上）参数使用。

要求：

（1）带单一参数时，参数名为 mapName；
（2）带两个参数时，参数名为 mapName，mapPosition；
（3）带三个参数时，参数名为 mapN，mapP，mapL。

2. 创建另外一个路由，名称为 OnlyTwoParametersAllows，这个方法只允许带有两个参数的路由通过，不允许带有多个参数或者不带参数的路由通过。

第 5 章

视 图

由于 MVC 将 IPO 相互隔离，因此将视图层单独分离出来，更有利于美工专心地处理界面的设计，程序员仅仅将页面的跳转逻辑设计好即可，这样可以使程序的开发更具高效性。本章内容更偏重于如何写好页面的跳转逻辑以及运用一些原本在 WebForm 里面的 HTML 标签，使程序员可以顺利地从 WebForm 过渡到 MVC 的设计逻辑之中。

5.1 视图概述

视图层（即 View 层）的主要作用是向用户提供用户界面，并负责和用户进行交互。它的动态数据主要源自于控制器层，而和用户交互后的数据又主要反馈给控制器层。它是用户和系统之间沟通的桥梁。

视图层中视图文件一般和控制器层中的有关函数存在着对应关系：控制器名称和 Views 文件夹（即视图层文件的根文件夹）下的子文件夹名称相对应；函数名称和子文件夹下面的视图文件名称相对应。一般在新建 MVC 4 工程后，会得到如图 5.1 所示的控制层文件，图中的控制器名称是 Home（字母 A），函数名称是 Index（字母 B）；相应的视图层文件结构如图 5.2 所示，可以看到 Views 文件夹下的 Home 文件夹（字母 A）和控制器名称相对应，视图文件 Index.cshtml（字母 B）和函数名称相对应。

```
public class HomeController : Controller
                 A
{
    public ActionResult Index()
                       B
    {
        ViewBag.Message = "修改此模板以快速启动你的 ASP.NET MVC 应用程序。";

        return View();
    }
```

图 5.1 控制层函数示意图

在 MVC 4 中，视图主要有 4 种：视图页、布局页、视图布局页、分部页。视图页一般在用户界面较为固定的情况下使用，页面不可拆分和复用。当存在一系列的用户界面，并且需要统一其中的部分界面时，一般对页面进行拆分，通过使用布局页和视图布局页来完成。布局页用于统一部分的界面布局；而视图布局页在"继承"布局页后，用于不同部分的界面布局。分部页一般用于精简结构、复用布局和局部刷新。

一般情况下，布局页（如图 5.2 中的_Layout.cshtml）放置在 Shared 文件夹（即公用视图文件夹）下，而其余 3 种视图（如图 5.2 中的 Index.cshtml）放置在控制器名称对应的文件夹下面。只有当视图需要跨控制器复用时，才需将视图放置在 Shared 文件夹下。

在 MVC 4 中，视图的引擎有两个，即 ASPX 引擎和 Razor 引擎。ASPX 引擎起源于微软公司的 WebForm 模式，在 MVC 和 MVC 2 中是唯一的视图引擎，但其在页面代码的编写方面存在诸多的不便。自 MVC 3 起，微软公司推出了 Razor 引擎，由于其编写的便利性，受到越来越多的程序员的喜爱。因此，本书将主要介绍 Razor 引擎下的视图。

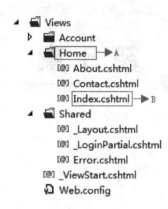

图 5.2 视图层文件结构图

5.2 视图页

5.2.1 视图页的创建

在 MVC 4 中，主要有以下 3 种方式来创建视图页。

1. 通过控制器中的函数快速创建

Step01：选中需要创建视图页的函数名称，如图 5.3 所示的 Start 函数。

Step02：在函数名称上单击右键，出现如图 5.4 所示的快捷菜单。

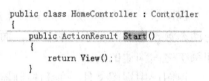

图 5.3 选中函数名称

图 5.4 右键快捷菜单

Step03：单击快捷菜单中的第一项（即"添加视图"），出现如图 5.5 所示的"添加视图"对话框。

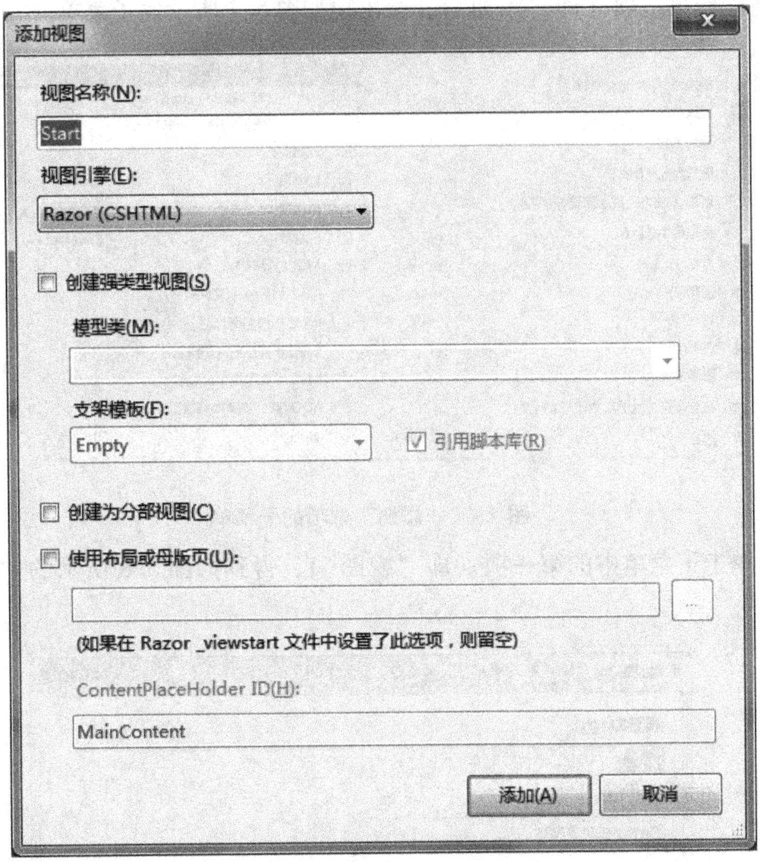

图 5.5 "添加视图"对话框

Step04：单击下方的"添加"按钮，即得到如图 5.6 所示的新创建的视图页页面。

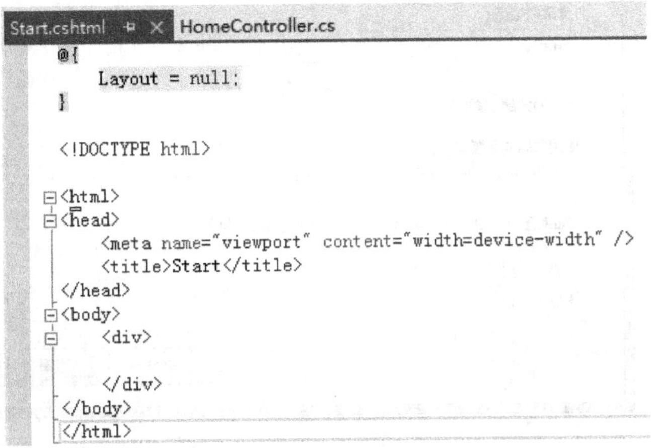

图 5.6 新创建的视图页页面

2. 通过"视图"创建

Step01：右键单击控制器名称表示的文件夹（本例中是 Home 文件夹），出现快捷菜单后，将鼠标滑动到第三项（即"添加"），会出现如图 5.7 所示的子菜单。

图 5.7 "添加"选项的子菜单

Step02：单击子菜单中的第一项（即"视图"），得到如图 5.8 所示的"添加视图"对话框。

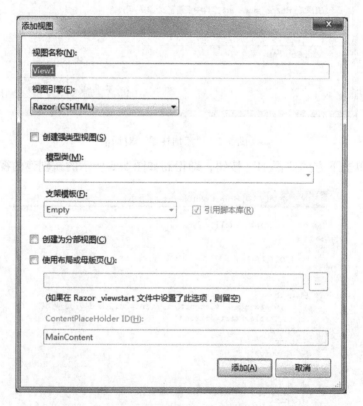

图 5.8 "添加视图"对话框

Step03：根据实际情况修改视图名称，本例中需将名字改为"Start"，修改后对话框状态和图 5.5 一致。

Step04：单击"添加"按钮，得到如图 5.6 所示的新创建的视图页页面。

3. 通过"新建项"创建

Step01：右键单击控制器名称表示的文件夹，出现快捷菜单，将鼠标滑动到第三项"添加"。

Step02：单击子菜单中的第三项（即"新建项"），得到如图 5.9 所示的"添加新项"对话框。

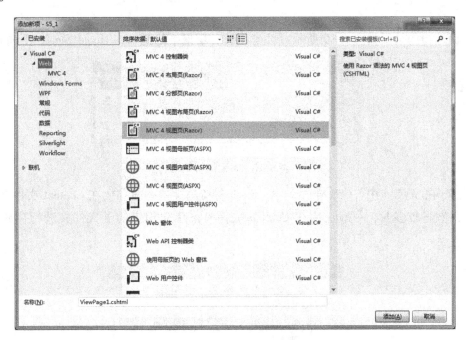

图 5.9 "添加新项"对话框

Step03：根据实际情况修改名称，本例中需将名称改为"Start.cshtml"。

Step04：单击下方的"添加"按钮，即得到如图 5.6 所示的新创建的视图页页面。

5.2.2 视图页介绍

在视图中，一般使用传统的 HTML 标记和 CSS 样式对页面布局进行控制，使用 JavaScript 和 JQuery 实现页面的动态效果，使用 C#代码进行后台数据的输出。本节将以举例的形式分别介绍这 3 种方法在视图中的使用。

1. HTML 标记和 CSS 样式

【例 5.1】视图中 HTML 标记和 CSS 样式的使用。

```
1    @{
2        Layout = null;
3    }
4    <!DOCTYPE html>
5    <html>
6    <head>
```

```
7        <meta name = "viewport" content = "width = device-width" />
8        <title>Start</title>
9     </head>
10    <body>
11       <div>
12           这是一个<span style = "color:red;font-weight:bold">开始</span>页面。
13       </div>
14    </body>
15 </html>
```

页面运行的结果如图 5.10 所示（注意网址，本例中的是 .../Home/Start，下同）。

图 5.10　页面运行结果

注意：在 MVC 4 中，CSS 样式文件、Flash 等资源文件一般存放在 Content 文件夹中；当资源文件数量较多时，一般还应在 Content 文件夹下建立相应的子文件夹进行存放（见图 5.11）。

图 5.11　Content 文件夹目录结构

2. JavaScript 和 JQuery

【例 5.2】视图中 JQuery 的使用。

```
1    @{
2        Layout = null;
3    }
4    <!DOCTYPE html>
5    <html>
6    <head>
7    <meta name = "viewport" content = "width = device-width" />
8    <title>Start</title>
9    <script src = "~/Scripts/jquery-1.7.1.js" type = "text/javascript"></script>
10   <script type = "text/javascript">
11          $(document).ready(function() {
```

```
12              var count = 0;
13              $("#btn").click(function () {
14                  count++;
15                  $("#btn").text("点击了" + count + "次！");
16              });
17          });
18      </script>
19  </head>
20  <body>
21      <div id="btn">
22          这是一个<span style="color:red;font-weight:bold">开始</span>页面。
23      </div>
24  </body>
25  </html>
```

页面运行的初始结果如图 5.10 所示。在点击文字若干次后，将得到类似如图 5.12 所示的文字。

图 5.12　多次点击后的运行结果

注意：在 MVC 4 中，JavaScript 和 JQuery 文件一般存放在 Scripts 文件夹中，如图 5.13 所示。

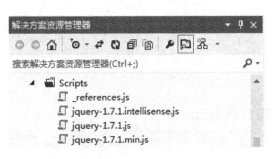

图 5.13　Scripts 文件夹目录结构

3. C#代码

在视图中编写 C#代码时，需要用@{}包裹，即@{代码段}；在输出有关变量时，需要用@引导，即@变量名；同时，在 MVC 4 中还支持 C#代码和 HTML 标记的混合编写。下面例子中的第一段 C#代码是纯 C#代码，第二段代码是混合编写的代码。

【例 5.3】视图中 C#代码的使用。

```
1   @{
2       Layout = null;
3   }

4   <!DOCTYPE html>
5   <html>
6   <head>
7       <meta name="viewport" content="width=device-width" />
8       <title>Start</title>
9   </head>
10  <body>
11      @{
12          for (int i = 0; i < 10; i++)
13          {
14              @i
15          }
16      }
17      @{
18          for (int i = 0; i < 5; i++)
19          {
20  <p>
21              @{
22                  for (int j = 0; j < i + 1; j++)
23                  {
24  <span>■</span>
25                  }
26              }
27  </p>
28          }
29      }
30  </body>
31  </html>
```

页面的运行结果如图 5.14 所示。

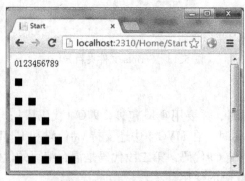

图 5.14 页面运行结果

注意：在C#代码段中，如果出现成对的HTML标记（如上例中的<p></p>），则这对标记之间的内容不认为被@{}包裹，编写代码段时需要重新用@{}包裹，如上例中关于j的循环段。上例中的第二段C#代码可以用下面例子中的代码等价替换。

【例5.4】等价的代码。

```
1    @{
2        for (int i = 0; i < 5; i ++)
3        {
4    <br />
5        for (int j = 0; j < i + 1; j ++)
6        {
7    <span>■</span>
8        }
9        }
10   }
```

5.3 从控制器层获取数据的方式

一般情况下，视图中的很多动态数据是需要从后台获取的。在MVC 4中，视图的数据是通过控制器层获取的，有弱类型、强类型、Session、Cookies这4种主要获取途径。后两者在各种网页开发技术中有着较为广泛的应用，而前两者是MVC 4中特有的。

5.3.1 弱类型

在MVC 4中，通过弱类型传递数据的方式有3种，即ViewBag、ViewData和TempData。

从生存周期来看，ViewBag和ViewData只有一个视图的周期，即在当前视图完全展示时，ViewBage和ViewData的数据将被清空，不能带入到下一个Action中；TempData的周期不定，一般在第一次使用后被清空，即如果在当前视图中未使用，则能带入到下一个Action中。

从数据类型来看，ViewData和TempData使用的是数据字典，在视图中使用时，一般需要进行数据类型的转换；ViewBag使用的是dynamic动态类型，在视图中可以直接使用，不用进行数据类型的转换。

下面通过几个例子来分别说明这三者之间的区别。

【例5.5】数据传递中ViewData和ViewBag的使用比较。

HomeController中的有关代码如下。

```
1    public ActionResult S1()
2    {
3        //生成初始列表
4        List<int> NumList = new List<int>();
5        for (int i = 0; i < 5; i ++)
6            NumList.Add(3 * i + 1);
7        //用不同方式传递
```

```
8            ViewData["List"] = NumList;
9            ViewBag.List = NumList;
10           return View();
11       }
```

S1.cshtml 中的代码如下。

```
1    @{
2        Layout = null;
3    }
4    <!DOCTYPE html>
5    <html>
6    <head>
7        <meta name="viewport" content="width=device-width" />
8        <title>S1</title>
9    </head>
10   <body>
11   <div>
12   使用 ViewData 获取数据：
13   <br />
14       @{
15           List<int> NumList1 = ViewData["List"] as List<int>;
16           for(int i = 0; i < NumList1.Count; i++)
17           {
18   <span>@NumList1[i] </span>
19           }
20       }
21   <br />
22   使用 ViewBag 获取数据：
23   <br />
24       @{
25           List<int> NumList2 = ViewBag.List;
26           for(int i = 0; i < NumList2.Count; i++)
27           {
28   <span>@NumList2[i] </span>
29           }
30       }
31   </div>
32   </body>
33   </html>
```

页面的运行结果如图 5.15 所示。

图 5.15　页面运行结果

【例 5.6】数据传递中 TempData 的使用。

HomeController 中的有关代码如下。

```
1       public ActionResult S2()
2       {
3           //初始数据
4           TempData["test"] = "TempData 有数据!";
5           return View();
6       }
7       public ActionResult S3()
8       {
9           //通过 ViewBag 接收传递过来的 TempData
10          ViewBag.test = TempData["test"];
11          return View();
12      }
```

S2.cshtml 中的代码如下。

```
1       @{
2           Layout = null;
3       }
4       <!DOCTYPE html>
5       <html>
6       <head>
7           <meta name="viewport" content="width=device-width" />
8           <title>S2</title>
9       </head>
10      <body>
11          @TempData["test"]
12          @{TempData.Keep("test");}
13          <br />
14          <a href="/Home/S3">点击测试跳转后是否有数据！</a>
15      </body>
16      </html>
```

S3.cshtml 中的代码如下。

```
1       @{
2           Layout = null;
```

```
3        }
4     <!DOCTYPE html>
5     <html>
6     <head>
7     <meta name = "viewport" content = "width = device-width" />
8     <title>S3</title>
9     </head>
10    <body>
11        @ViewBag.test
12    </body>
13    </html>
```

页面的初始运行结果如图 5.16 所示。单击超链接后，页面的运行结果如图 5.17 所示。可以看到，TempData 中的数据依然存在，这是因为 S2.cshtml 中使用了 @{TempData.Keep("test");} 对 TempData 中的数据进行了保留；如果删掉这段代码，页面的运行结果将如图 5.18 所示。

在视图中使用 TempData 后，若想继续保留，则可以通过 TempData.Keep("关键字名称")这一函数实现。

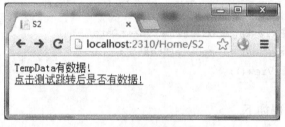

图 5.16　页面初始运行结果

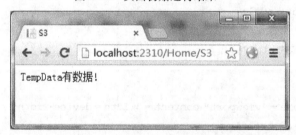

图 5.17　单击超链接后的运行结果

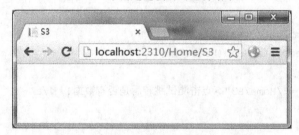

图 5.18　去掉部分代码后的运行结果

注意：当 ViewData 和 TempData 中的数据作为字符串类型使用时，可以不进行数据类型的转换。

5.3.2 强类型

在 MVC 4 中，主要是通过 Model 来实现强类型的数据传递，需要在视图页面上声明 Model 的具体类型，可以通过两种方式实现：一种是在创建视图的时候，勾选"创建强类型视图"复选框，并选择模型类（见图 5.19）；另一种是在创建普通的视图后，在顶端用"@model 模型类"声明（见下文中的例子）。

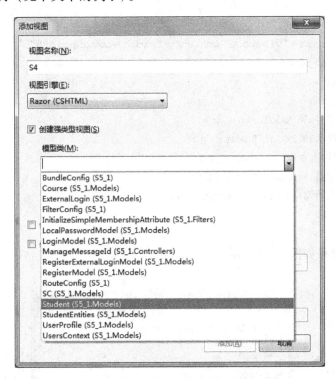

图 5.19 创建带强类型的视图

在控制器层，一般通过 return View（数据名称）进行传递；在视图层，一般通过 model 这个专用变量进行使用。具体使用方法可参考下面的例子。

【例 5.7】数据传递中强类型的使用。

在顶端引用模型：using S5 _ 1. Models。

有关 Action：为了举例的方便，本例中将数据的读取放在了控制器层，建议读者在实际操作时将其放在模型层。

HomeController 中的有关代码如下。

```
1    public ActionResult S4()
2    {
3        //获取数据库连接
4        StudentEntities db = new StudentEntities();
5        //读取第一个学生的信息
6        Student s1 = db.Students.FirstOrDefault();
7        return View(s1);
8    }
```

S4.cshtml 中的代码如下。

```
1      @ model S5_1.Models.Student
2      @{
3          Layout = null;
4      }
5      <!DOCTYPE html>
6      <html>
7      <head>
8          <meta name = "viewport" content = "width = device-width" />
9          <title>S4</title>
10     </head>
11     <body>
12         <p>学号:@Model.Sno</p>
13         <p>姓名:@Model.Sname</p>
14         <p>性别:@Model.Ssex</p>
15         <p>年龄:@Model.Sage</p>
16         <p>系所缩写:@Model.Sdept</p>
17     </body>
18     </html>
```

页面的运行结果如图 5.20 所示。

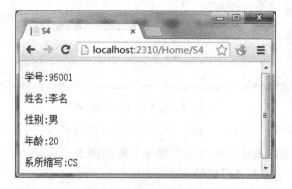

图 5.20　页面运行结果

注意：@ model 可以使用 Model 的复合类型，如 List < Model >（见例 5.8）。此外，当视图中使用的 Model 多于 1 个时，其余的 Model 需要通过"@ using 模型类"来声明，并且只能通过弱类型的方式来传递（见例 5.9）。

【例 5.8】数据传递中复合 Model 类型的使用。

HomeController 中的有关代码如下。

```
1      public ActionResult S4()
2      {
3          //获取数据库连接
4          StudentEntities db = new StudentEntities();
```

```
5          //读取学生的信息
6          List<Student> ss=db.Students.ToList();
7          return View(ss);
8      }
```

S4.cshtml 中的代码如下。

```
1   @model List<S5_1.Models.Student>
2   @{
3       Layout = null;
4   }
5   <!DOCTYPE html>
6   <html>
7   <head>
8       <meta name="viewport" content="width=device-width" />
9       <title>S4</title>
10  </head>
11  <body>
12      <table>
13      <tr>
14      <td>学号</td>
15      <td>姓名</td>
16      <td>性别</td>
17      <td>年龄</td>
18      <td>系所缩写</td>
19      </tr>
20          @{
21              for (int i=0; i<Model.Count; i++)
22              {
23      <tr>
24      <td>@Model[i].Sno</td>
25      <td>@Model[i].Sname</td>
26      <td>@Model[i].Ssex</td>
27      <td>@Model[i].Sage</td>
28      <td>@Model[i].Sdept</td>
29      </tr>
30              }
31          }
32  </body>
33  </html>
```

页面的运行结果如图 5.21 所示。

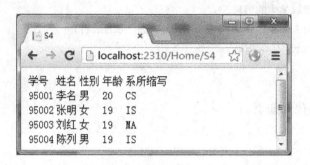

图 5.21　页面运行结果

【例 5.9】数据传递中多 Model 的使用。

HomeController 中的有关代码如下。

```
1      public ActionResult S4()
2      {
3          //获取数据库连接
4          StudentEntities db = new StudentEntities();
5          //读取第一个学生的信息
6          Student s1 = db.Students.FirstOrDefault();
7          //读取第一个课程的信息
8          Course c1 = db.Courses.FirstOrDefault();
9          ViewBag.course = c1;
10         return View(s1);
11     }
```

S4.cshtml 中的代码如下。

```
1      @model S5_1.Models.Student
2      @using S5_1.Models;
3      @{
4          Layout = null;
5      }
6      <!DOCTYPE html>
7      <html>
8      <head>
9      <meta name="viewport" content="width=device-width" />
10     <title>S4</title>
11     </head>
12     <body>
13     <table>
14     <tr>
15     <td>学生信息:
16     <p>学号:@Model.Sno</p>
17     <p>姓名:@Model.Sname</p>
18     <p>性别:@Model.Ssex</p>
```

```
19      <p>年龄:@Model.Sage</p>
20      </td>
21      <td width = "20px"></td>
22      <td>课程信息:
23              @{Course c = ViewBag.course;}
24      <p>编号:@c.Cno</p>
25      <p>名称:@c.Cname</p>
26      <p>学分:@c.Ccredit</p>
27      <p>先行课程编号:@c.Cpno</p>
28      </td>
29      </tr>
30      </table>
31      </body>
32      </html>
```

页面的运行结果如图5.22所示。

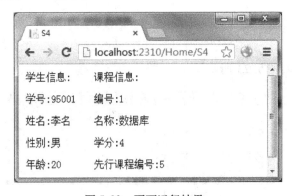

图5.22 页面运行结果

5.3.3 Session 和 Cookies

在 MVC 4 中,可以通过 Session 这个关键字来实现 Session 的功能。Session 的使用方式和 ViewData、TempData 类似;可以通过 Response.Cookies["关键字"].Value 来实现 Cookies 的写入,通过 Request.Cookies["关键字"].Value 来实现 Cookies 的读取。Session 和 Cookies 的生存周期都以时间来度量,可以根据实际情况调整。下面的例5.10展示了它们在 MVC 4 中的使用。

【例5.10】数据传递中 Session 和 Cookies 的使用。

HomeController 中的有关代码如下。

```
1   public ActionResult S5()
2   {
3       Session["data"] = "这是通过 Session 传递的数据!";
4       Response.Cookies["data"].Value = "这是通过 Cookies 传递的数据!";
5       return View();
6   }
```

S5.cshtml 中的代码如下。

```
1       @{
2           Layout = null;
3       }
4       <!DOCTYPE html>
5       <html>
6       <head>
7           <meta name="viewport" content="width=device-width" />
8           <title>S5</title>
9       </head>
10      <body>
11          <div>
12              @Session["data"]
13              <br />
14              @Request.Cookies["data"].Value
15          </div>
16      </body>
17      </html>
```

页面的运行结果如图 5.23 所示。

图 5.23 页面运行结果

5.4 HtmlHelper 类

为了方便视图层代码的编写，微软公司推出了 HtmlHelper 类，用于帮助开发人员快速生成 HTML 代码。其中，常见的方法及其用途见表 5.1。下面将分别对这些方法进行介绍。

表 5.1 常见方法及其用途

序号	方法	用途	对应的 HTML 标记
1	ActionLink	生成超链接	<a></a>
2	BeginForm	生成表单	<form>
3	EndForm		</form>
4	CheckBox	生成复选框	<input type="checkbox"/>
5	DropDownList	生成下拉框	<select><option></option></select>
6	Hidden	生成隐藏输入框	<input type="hidden"/>

（续）

序号	方法	用途	对应的 HTML 标记
7	Label	生成文本	< label > < /label >
8	ListBox	生成列表框	< select multiple = "multiple" > < option > < /option > < /select >
9	Password	生成密码输入框	< input type = "password"/ >
10	RadioButton	生成单选框	< input type = "radio"/ >
11	TextArea	生成多行文本输入框	< textarea > < /textarea >
12	TextBox	生成文本输入框	< input type = "text"/ >

5.4.1 ActionLink

ActionLink 方法共有 10 种重载方式，下面的 5 种较为常用。

1）ActionLink（string linkText, string actionName）
2）ActionLink（string linkText, string actionName, object routeValues）
3）ActionLink（string linkText, string actionName, string controllerName）
4）ActionLink（string linkText, string actionName, object routeValues, object htmlAttributes）
5）ActionLink（string linkText, string actionName, string controllerName, object routeValues, object htmlAttributes）

各参数的作用如下。

1）linkText：设置超链接的显示文本。
2）actionName：设置超链接的目标 Action。
3）controllerName：设置超链接的目标 Controller；当方法中无该参数时，默认和当前视图同一个 Controller。
4）routeValues：设置超链接通过路由传递的参数。
5）htmlAttributes：设置超链接的 HTML 属性。

下面通过一个例子来看一下这 5 种方法的具体使用情况。

【例 5.11】常见的 ActionLink 用法。

```
1       @ {
2           Layout = null;
3       }
4       <! DOCTYPE html >
5       < html >
6       < head >
7       < meta name = "viewport" content = "width = device-width" / >
8       < title > S6 < /title >
9       < /head >
10      < body >
11      < div >
12              @ Html.ActionLink("方法(1)", "S6")
```

```
13      @Html.ActionLink("方法(2)", "S6", new { value = 1, name = "s6" })
14      @Html.ActionLink("方法(3)", "S6", "Home")
15      @Html.ActionLink("方法(4)", "S6", new { id = 1 }, new { style = "font-
16   weight:bold" })
17       @Html.ActionLink("方法(5)", "S6", "Home", new { id = 1 }, new
18   { style = "font-weight:bold" })
19   </div>
20   </body>
21   </html>
```

页面的运行结果如图 5.24 所示。

图 5.24 页面运行结果

查看源文件后，可得到如下的 HTML 代码。

```
1    <a href="/Home/S6">方法(1)</a>
2    <a href="/Home/S6?value=1&name=s6">方法(2)</a>
3    <a href="/Home/S6">方法(3)</a>
4    <a href="/Home/S6/1" style="font-weight:bold">方法(4)</a>
5    <a href="/Home/S6/1" style="font-weight:bold">方法(5)</a>
```

5.4.2 BeginForm 和 EndForm

BeginForm 方法共有 13 种重载方式，下面的 6 种较为常用。

1) BeginForm (string actionName, string controllerName)

2) BeginForm (string actionName, string controllerName, FormMethod method)

3) BeginForm (string actionName, string controllerName, object routeValues)

4) BeginForm (string actionName, string controllerName, FormMethod method, object htmlAttributes)

5) BeginForm (string actionName, string controllerName, object routeValues, FormMethod method)

6) BeginForm (string actionName, string controllerName, object routeValues, FormMethod method, object htmlAttributes)

各参数的作用如下。

1) actionName：设置表单提交后的目标 Action。

2) controllerName：设置表单提交后的目标 Controller。

3) method：设置表单的提交方式（Post 方式或 Get 方式）；方法中无该参数时，默认使用 Post 方式。

4）routeValues：设置表单通过路由传递的参数。

5）htmlAttributes：设置表单的 HTML 属性。

下面通过一个例子来看一下这 6 种方法的具体使用情况。

【例 5.12】常见的 BeginForm 用法。

```
1   @{
2       Layout = null;
3   }
4   <!DOCTYPE html>
5   <html>
6   <head>
7   <meta name="viewport" content="width=device-width" />
8   <title>S7</title>
9   </head>
10  <body>
11      @using (Html.BeginForm("S7", "Home"))
12      {
13  <li>方法(1)</li>
14      }
15      @using (Html.BeginForm("S7", "Home",FormMethod.Get))
16      {
17  <li>方法(2)</li>
18      }
19      @using (Html.BeginForm("S7", "Home", new { id=1 }))
20      {
21  <li>方法(3)</li>
22      }
23      @using (Html.BeginForm("S7", "Home", FormMethod.Get,new { name="S7" }))
24      {
25  <li>方法(4)</li>
26      }
27      @using (Html.BeginForm("S7", "Home", new { id=1 }, FormMethod.Get))
28      {
29  <li>方法(5)</li>
30      }
31      @using (Html.BeginForm("S7", "Home", new { id=1 }, FormMethod.Get, new {
32  name="S7" }))
33      {
34  <li>方法(6)</li>
35      }
36  </body>
37  </html>
```

页面运行后，查看源文件可得到如下的 HTML 代码。

```
1    <form action="/Home/S7" method="post"><li>方法(1)</li></form>
2    <form action="/Home/S7" method="get"><li>方法(2)</li></form>
3    <form action="/Home/S7/1" method="post"><li>方法(3)</li></form>
4    <form action="/Home/S7" method="get" name="S7"><li>方法(4)</li></form>
5    <form action="/Home/S7/1" method="get"><li>方法(5)</li></form>
6    <form action="/Home/S7/1" method="get" name="S7"><li>方法(6)</li></form>
```

EndForm 方法无重载方法，只有 EndForm( )方式。

在生成表单时，有两种方法：一种是仅利用 BeginForm( )，在函数后面用{ }把表单内容包裹起来（见例 5.12）；另一种是利用 BeginForm( ) 和 EndForm( ) 分别生成 <form> 和 </form>，将表单内容包裹在中间（见例 5.13）。

【例 5.13】BeginForm 和 EndForm 的用法。

```
1    @{
2        Layout = null;
3    }
4    <!DOCTYPE html>
5    <html>
6    <head>
7    <meta name="viewport" content="width=device-width"/>
8    <title>S8</title>
9    </head>
10   <body>
11       @{Html.BeginForm("S8", "Home", new { id = 1 }, FormMethod.Get, new
12   { name = "S7" });}
13   这是包裹在 BeginForm 和 EndForm 之间的表单内容
14       @{Html.EndForm();}
15   </body>
16   </html>
```

页面运行后，查看源文件可得到如下的 HTML 代码。

```
1    <form action="/Home/S7/1" method="get" name="S8">
2    这是包裹在 BeginForm 和 EndForm 之间的表单内容
3    </form>
```

5.4.3 CheckBox

CheckBox 方法共有 6 种重载方式，下面的 4 种较为常用。

1) CheckBox (string name)
2) CheckBox (string name, bool isChecked)
3) CheckBox (string name, object htmlAttributes)
4) CheckBox (string name, bool isChecked, object htmlAttributes)

各参数的作用如下。

1) name：设置复选框的名字及 ID。
2) isChecked：设置复选框的初始选中情况，true 是选中，false 是不选中；方法中无该参数时，默认不选中。

3) htmlAttributes：设置复选框的 HTML 属性。

下面通过一个例子来看一下这 4 种方法的具体使用情况。

【例 5.14】常见的 CheckBox 用法。

```
1    @{
2        Layout = null;
3    }
4    <!DOCTYPE html>
5    <html>
6    <head>
7    <meta name="viewport" content="width=device-width" />
8    <title>S9</title>
9    </head>
10   <body>
11   <div>
12        @Html.CheckBox("S9")方法(1)
13        @Html.CheckBox("S9",true)方法(2)
14        @Html.CheckBox("S9", new { id = "S" })方法(3)
15        @Html.CheckBox("S9", true, new { id = "S" })方法(4)
16   </div>
17   </body>
18   </html>
```

页面的运行结果如图 5.25 所示。

图 5.25 页面运行结果

查看源文件后，可得到如下的 HTML 代码。

```
1    <input id="S9" name="S9" type="checkbox" value="true" /><input name="
2    S9" type="hidden" value="false" />方法(1)
3    <input checked="checked" id="S9" name="S9" type="checkbox" value="
4    true" /><input name="S9" type="hidden" value="false" />方法(2)
5    <input id="S" name="S9" type="checkbox" value="true" /><input name="
6    S9" type="hidden" value="false" />方法(3)
7    <input checked="checked" id="S" name="S9" type="checkbox" value="true"
8    /><input name="S9" type="hidden" value="false" />方法(4)
```

5.4.4 DropDownList

DropDownList 方法共有 8 种重载方式，下面的 4 种较为常用。

1）DropDownList（string name，IEnumerable < SelectListItem > selectList）

2）DropDownList（string name，IEnumerable < SelectListItem > selectList，object htmlAttributes）

3）DropDownList（string name，IEnumerable < SelectListItem > selectList，string optionLabel）

4）DropDownList（string name，IEnumerable < SelectListItem > selectList，string optionLabel，object htmlAttributes）

各参数的作用如下。

1）name：设置下拉框的名字及 ID。

2）selectList：设置下拉框的选项内容，一般在控制器层生成。

3）htmlAttributes：设置下拉框的 HTML 属性。

4）optionLabel：设置初始状态下下拉框显示的内容。

下面通过一个例子来看一下这 4 种方法的具体使用情况。

【例 5.15】常见的 DropDownList 用法。

HomeController 中的有关代码如下。

```
1    public ActionResult S10()
2    {
3        List < SelectListItem > list = new List < SelectListItem > ();
4        for (int i = 0; i < 5; i ++)
5        list.Add(new SelectListItem { Text = "选项" + (i + 1), Value = (i + 1).ToS-
6        tring() });
7        ViewBag.list = list;
8        return View();
9    }
```

S10.cshtml 中的代码如下。

```
1    @{
2        Layout = null;
3    }
4    <! DOCTYPE html >
5    < html >
6    < head >
7    < meta name = "viewport" content = "width = device-width" />
8    < title > S10 </title >
9    </head >
10   < body >
11   < div >
12   方法（1）：@ Html.DropDownList（" S10 "，ViewBag.list as IEnumerable
13   < SelectListItem >）
14   方法（2）：@ Html.DropDownList（" S10 "，ViewBag.list as IEnumerable
15   < SelectListItem >，new { style = "width:100px" }）
16   方法（3）：@ Html.DropDownList（" S10 "，ViewBag.list as IEnumerable
17   < SelectListItem >，"初始选项"）
```

```
18        方法(4)：@ Html.DropDownList (" S10 ", ViewBag.list as IEnumerable
19        <SelectListItem>,"初始选项", new { style = "width:100px" })
20     </div>
21   </body>
22 </html>
```

页面的运行结果如图 5.26 所示。

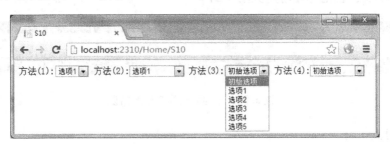

图 5.26　页面运行结果

查看源文件后，可得到如下的 HTML 代码。

```
1    方法(1)：< select id = "S10" name = "S10" >
2    < option value = "1" >选项 1 </option >
3    < option value = "2" >选项 2 </option >
4    < option value = "3" >选项 3 </option >
5    < option value = "4" >选项 4 </option >
6    < option value = "5" >选项 5 </option >
7    </select >
8    方法(2)：< select id = "S10" name = "S10" style = "width:100px" >
9    < option value = "1" >选项 1 </option >
10   < option value = "2" >选项 2 </option >
11   < option value = "3" >选项 3 </option >
12   < option value = "4" >选项 4 </option >
13   < option value = "5" >选项 5 </option >
14   </select >
15   方法(3)：< select id = "S10" name = "S10" >
16   < option value = "" >初始选项 </option >
17   < option value = "1" >选项 1 </option >
18   < option value = "2" >选项 2 </option >
19   < option value = "3" >选项 3 </option >
20   < option value = "4" >选项 4 </option >
21   < option value = "5" >选项 5 </option >
22   </select >
23   方法(4)：< select id = "S10" name = "S10" style = "width:100px" >
24   < option value = "" >初始选项 </option >
25   < option value = "1" >选项 1 </option >
26   < option value = "2" >选项 2 </option >
```

```
27        < option value = "3" > 选项 3 < /option >
28        < option value = "4" > 选项 4 < /option >
29        < option value = "5" > 选项 5 < /option >
30    </select >
```

注意：通过 ViewBag 等弱类型传递过来的选项列表需要强制转换为 IEnumerable < SelectListItem >，否则将报错（见图 5.27）。

图 5.27 未强制转换产生的错误

5.4.5 Hidden

Hidden 方法共有 4 种重载方式，下面的 3 种较为常用。

1) Hidden (string name)
2) Hidden (string name, object value)
3) Hidden (string name, object value, object htmlAttributes)

各参数的作用如下。

1) name：设置隐藏输入框的名字和 ID。
2) value：设置隐藏输入框的值。
3) htmlAttributes：设置隐藏输入框的 HTML 属性。

下面通过一个例子来看一下这 3 种方法的具体使用情况。

【例 5.16】常见的 Hidden 用法。

```
1     @ {
2         Layout = null;
3     }
4     <! DOCTYPE html >
5     < html >
6     < head >
7     < meta name = "viewport" content = "width = device-width" / >
8     < title > S11 </title >
9     </head >
10    < body >
11    < div >
12    方法(1):@ Html.Hidden("S11")
13    方法(2):@ Html.Hidden("S11","数据值")
14    方法(3):@ Html.Hidden("S11", "数据值", new { id = "S" })
15    </div >
16    </body >
17    </html >
```

页面运行后,查看源文件可得到如下的 HTML 代码。

1   方法(1):< input id = "S11" name = "S11" type = "hidden" value = "" / >
2   方法(2):< input id = "S11" name = "S11" type = "hidden" value = "数据值" / >
3   方法(3):< input id = "S" name = "S11" type = "hidden" value = "数据值" / >

5.4.6 Label

Label 方法共有 6 种重载方式,下面的两种较为常用。

1) Label(string expression)
2) Label(string expression, object htmlAttributes)

各参数的作用如下。

1) expression:设置文本的内容。
2) htmlAttributes:设置文本的 HTML 属性。

下面通过一个例子来看一下这两种方法的具体使用情况。

【例 5.17】 常见的 Label 用法。

```
1    @ {
2        Layout = null;
3    }
4    <! DOCTYPE html >
5    < html >
6    < head >
7    < meta name = "viewport" content = "width = device-width" / >
8    < title > S12 </ title >
9    </ head >
10   < body >
11   < div >
12       @ Html.Label("方法(1)")
13       @ Html.Label("方法(2)", new { style = "color: red;font-weight:bold;
14   font-style: italic" })
15   </ div >
16   </ body >
17   </ html >
```

页面的运行结果如图 5.28 所示。

图 5.28　页面运行结果

查看源文件后,可得到如下的 HTML 代码。

1   < label for = "" >方法(1) </ label >
2   < label for = "" style = "color: red;font-weight:bold;font-style: italic" >方法(2)
3   </ label >

5.4.7 ListBox

ListBox 方法共有 4 种重载方式,下面的两种较为常用。
1) ListBox(string name, IEnumerable < SelectListItem > selectList)
2) ListBox(string name, IEnumerable < SelectListItem > selectList, object htmlAttributes)
各参数的作用如下。
1) name:设置列表框的名字及 ID。
2) selectList:设置列表框的选项内容,一般在控制器层生成。
3) htmlAttributes:设置列表框的 HTML 属性。
下面通过一个例子来看一下这两种方法的具体使用情况。

【例 5.18】 常见的 ListBox 用法。

HomeController 中的有关代码如下。

```
1    public ActionResult S13()
2    {
3        List < SelectListItem > list = new List < SelectListItem > ();
4        for (int i = 0; i < 5; i ++ )
5            list.Add(new SelectListItem { Text = "选项" + (i + 1), Value = (i + 1).ToS-
6    tring() });
7        ViewBag.list = list;
8        return View();
9    }
```

S13.cshtml 中的代码如下。

```
1    @{
2        Layout = null;
3    }
4    <! DOCTYPE html >
5    < html >
6    < head >
7    < meta name = "viewport" content = "width = device-width" / >
8    < title > S13 </title >
9    </head >
10   < body >
11   < div >
12   方法(1):@Html.ListBox("S12", ViewBag.list as IEnumerable < SelectListItem > )
13   方法(2):@Html.ListBox("S12", ViewBag.list as IEnumerable < SelectListItem > , new {
14   style = "width:100px" })
15   </div >
16   </body >
17   </html >
```

页面的运行结果如图 5.29 所示。

图 5.29　页面运行结果

查看源文件后，可得到如下的 HTML 代码。

```
1    方法(1)：<select id = "S12" multiple = "multiple" name = "S12" >
2      <option value = "1" >选项1</option>
3      <option value = "2" >选项2</option>
4      <option value = "3" >选项3</option>
5      <option value = "4" >选项4</option>
6      <option value = "5" >选项5</option>
7    </select>
8    方法(2)：<select id = "S12" multiple = "multiple" name = "S12" style = "width:
9      100px" >
10     <option value = "1" >选项1</option>
11     <option value = "2" >选项2</option>
12     <option value = "3" >选项3</option>
13     <option value = "4" >选项4</option>
14     <option value = "5" >选项5</option>
15   </select>
```

注意：通过 ViewBag 等弱类型传递过来的选项列表需要强制转换为 IEnumerable <SelectListItem>，否则将报错。

5.4.8　Password

Password 方法共有 4 种重载方式，下面的两种较为常用。

1）Password（string name）

2）Password（string name, object value, object htmlAttributes）

各参数的作用如下。

1）name：设置密码输入框的名字和 ID。

2）value：设置密码输入框的值，一般设置为空。

3）htmlAttributes：设置密码输入框的 HTML 属性。

下面通过一个例子来看一下这两种方法的具体使用情况。

【例 5.19】常见的 Password 用法。

```
1    @{
2        Layout = null;
3    }
4    <! DOCTYPE html >
```

```
5       <html>
6       <head>
7       <meta name="viewport" content="width=device-width" />
8       <title>S14</title>
9       </head>
10      <body>
11      <div>
12      方法(1):@Html.Password("S14")
13      方法(2):@Html.Password("S14", "123456", new { style = "width:100px" })
14      </div>
15      </body>
16      </html>
```

页面的运行结果如图 5.30 所示。

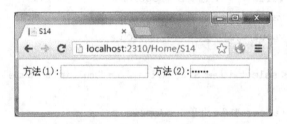

图 5.30 页面运行结果

查看源文件后,可得到如下的 HTML 代码。

```
1       方法(1):<input id="S14" name="S14" type="password" />
2       方法(2):<input id="S14" name="S14" style="width:100px" type="password" value
3       ="123456" />
```

5.4.9 RadioButton

RadioButton 方法共有 6 种重载方式,下面的 4 种较为常用。

1) RadioButton (string name, object value)
2) RadioButton (string name, object value, bool isChecked)
3) RadioButton (string name, object value, object htmlAttributes)
4) RadioButton (string name, object value, bool isChecked, object htmlAttributes)

各参数的作用如下。

1) name:设置单选框的名字及 ID。
2) value:设置单选框的值。
3) isChecked:设置单选框的初始选中情况,true 是选中,false 是不选中;当方法中无该参数时,默认不选中。
4) htmlAttributes:设置单选框的 HTML 属性。

下面通过一个例子来看一下这 4 种方法的具体使用情况。

【例 5.20】 常见的 RadioButton 用法。

```
1     @{
2         Layout = null;
3     }
4     <!DOCTYPE html>
5     <html>
6     <head>
7         <meta name="viewport" content="width=device-width" />
8         <title>S15</title>
9     </head>
10    <body>
11        <div>
12            @Html.RadioButton("S15","数据值")方法(1)
13            @Html.RadioButton("S15","数据值",true)方法(2)
14            @Html.RadioButton("S15_2","数据值", new { id = "S" })方法(3)
15            @Html.RadioButton("S15_2","数据值", true, new { id = "SS" })方法(4)
16        </div>
17    </body>
18    </html>
```

页面的运行结果如图 5.31 所示。

图 5.31　页面运行结果

查看源文件后，可得到如下的 HTML 代码。

```
1    <input id="S15" name="S15" type="radio" value="数据值" />方法(1)
2    <input checked="checked" id="S15" name="S15" type="radio" value="数据
3    值" />方法(2)
4    <input id="S" name="S15_2" type="radio" value="数据值" />方法(3)
5    <input checked="checked" id="SS" name="S15_2" type="radio" value="数据
6    值" />方法(4)
```

注意：具有相同 name 属性的单选框，只有一个能被选中。

5.4.10　TextArea

TextArea 方法共有 8 种重载方式，下面的 5 种较为常用。

1）TextArea（string name）
2）TextArea（string name, object htmlAttributes）
3）TextArea（string name, string value）
4）TextArea（string name, string value, object htmlAttributes）
5）TextArea（string name, string value, int rows, int columns, object htmlAttributes）

各参数的作用如下。
1) name：设置多行文本输入框的名字及ID。
2) value：设置多行文本输入框的值。
3) htmlAttributes：设置多行文本输入框的HTML属性。
4) rows：设置多行文本输入框的行数。
5) columns：设置多行文本输入框的列数。

下面通过一个例子来看一下这5种方法的具体使用情况。

【例5.21】 常见的TextArea用法。

```
1    @{
2        Layout = null;
3    }
4    <!DOCTYPE html>
5    <html>
6    <head>
7    <meta name="viewport" content="width=device-width"/>
8    <title>S16</title>
9    </head>
10   <body>
11   <div>
12   方法(1):@Html.TextArea("S16")
13   方法(2):@Html.TextArea("S16", new { @style = "width:100px;height:50px" })
14       @Html.TextArea("S16","方法(3)")
15       @Html.TextArea("S16","方法(4)", new { @style = "width:100px;
16   height:50px" })
17       @Html.TextArea("S16", "方法(5)", 5, 20, new { @id = "S" })
18   </div>
19   </body>
20   </html>
```

页面的运行结果如图5.32所示。

图5.32 页面运行结果

查看源文件后，可得到如下的HTML代码。

```
1    方法(1):<textarea cols="20" id="S16" name="S16" rows="2"></textarea>
2    方法(2):<textarea cols="20" id="S16" name="S16" rows="2"
3    style="width:100px;height:50px">
4    </textarea>
```

```
5       <textarea cols="20" id="S16" name="S16" rows="2">方法(3)</textarea>
6       <textarea cols="20" id="S16" name="S16" rows="2" style="width:100px;
7       height:50px">方法(4)
8       </textarea>
9       <textarea cols="20" id="S" name="S16" rows="5">方法(5)</textarea>
```

5.4.11　TextBox

TextBox 方法共有 7 种重载方式，下面的 3 种较为常用。

1）TextBox（string name）
2）TextBox（string name，string value）
3）TextBox（string name，string value，object htmlAttributes）

各参数的作用如下。

1）name：设置文本输入框的名字及 ID。
2）value：设置文本输入框的值。
3）htmlAttributes：设置文本输入框的 HTML 属性。

下面通过一个例子来看一下这 3 种方法的具体使用情况。

【例 5.22】常见的 TextBox 用法。

```
1       @{
2           Layout = null;
3       }
4       <!DOCTYPE html>
5       <html>
6       <head>
7       <meta name="viewport" content="width=device-width" />
8       <title>S17</title>
9       </head>
10      <body>
11      <div>
12      方法(1):@Html.TextBox("S17")
13      方法(2):@Html.TextBox("S17","方法(2)")
14      方法(3):@Html.TextBox("S17","方法(3)", new { style="width:100px" })
15      </div>
16      </body>
17      </html>
```

页面的运行结果如图 5.33 所示。

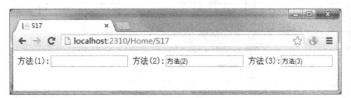

图 5.33　页面运行结果

查看源文件后,可得到如下的 HTML 代码。

```
1    方法(1):<input id="S17" name="S17" type="text" value="" />
2    方法(2):<input id="S17" name="S17" type="text" value="方法(2)" />
3    方法(3):<input id="S17" name="S17" style="width:100px" type="text" value
4    ="方法(3)" />
```

5.5 布局页和视图布局页

5.5.1 布局页的创建

在 MVC 4 中,布局页一般通过"新建项"来创建。

Step01:右键单击 Shared 文件夹,出现快捷菜单后,将鼠标滑动到第三项(即"添加"),会出现如图 5.34 所示的子菜单。

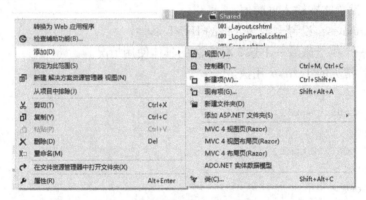

图 5.34 快捷菜单中的子菜单

Step02:单击子菜单中的第三项(即"新建项"),得到如图 5.35 所示的"添加新项"对话框。

图 5.35 "添加新项"对话框

Step03：根据实际情况修改名称，本例中将名称改为"_MyLayoutPage.cshtml"。
Step04：单击下方的"添加"按钮，即得到如图5.36所示的新创建的布局页页面。

图5.36　新创建的布局页页面

5.5.2　视图布局页的创建

在MVC 4中，视图布局页的创建过程与视图页的创建过程类似，也可以通过以下3种方式来创建。

1. 通过控制器中的函数快速创建

Step01 ~ Step03：同"通过控制器中的函数快速创建"视图页过程（见5.2.1节）中的Step01 ~ Step03。

Step04：勾选"使用布局或母版页"复选框，如图5.37所示。

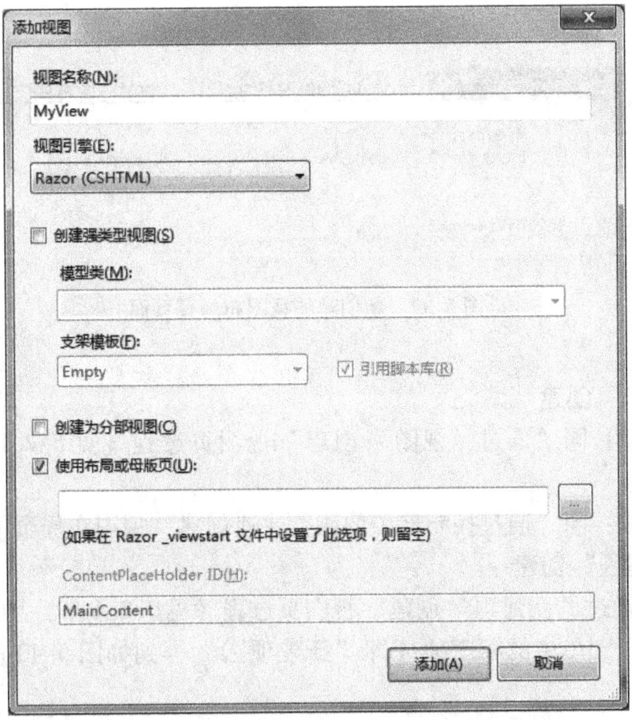

图5.37　勾选"使用布局或母版页"复选框

Step05：单击"…"按钮，出现如图5.38所示的"选择布局页"对话框。

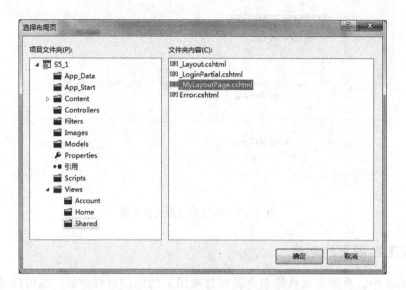

图5.38 "选择布局页"对话框

Step06：在Shared文件下选中需要的布局页之后，单击"确定"按钮。

Step07：在返回"添加视图"对话框后，单击下方的"添加"按钮，即得到如图5.39所示的新创建的视图布局页页面。

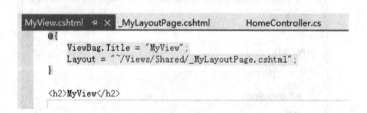

图5.39 新创建的视图布局页页面

2. 通过"视图"创建

Step01～Step03：同"通过'视图'创建"视图页过程（见5.2.1节）中的Step01～Step03。

Step04～Step07：同"通过控制器中的函数快速创建"视图布局页中的Step04～Step07。

3. 通过"新建项"创建

Step01：同"通过'新建项'创建"视图页过程（见5.2.1节）中的Step01。

Step02：单击子菜单中的第三项（即"新建项"），得到如图5.40所示的"添加新项"对话框。

第 5 章 视　图　133

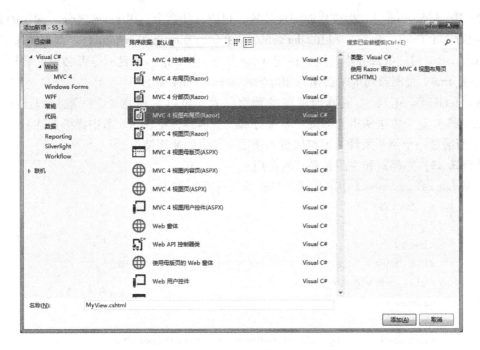

图 5.40　"添加新项"对话框

Step03：根据实际情况修改名称，本例中需将名称改为"MyView.cshtml"。
Step04：单击下方的"添加"按钮，出现如图 5.38 所示的"选择布局页"对话框。
Step05：在 Shared 文件下选中需要的布局页之后，单击"确定"按钮，即得到如图 5.39 所示的新创建的视图布局页页面。

5.5.3　布局页和视图布局页介绍

布局页和视图布局页一般配合使用，其关系示意图如图 5.41 所示。布局页相当于一个模板，在进行总体布局时，会预留若干位置，如图 5.41 中的"预留位置 1"和"预留位置 2"。视图布局页在"继承"视图页后，只需对预留位置进行布局（如图 5.41 中的"视图代码段 1"和"视图代码段 2"），即可得到完整的视图。

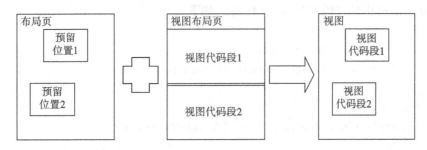

图 5.41　布局页和视图布局页之间的关系示意图

在布局页，主要通过 3 种方式来预留位置。
1）RenderBody()：生成主要预留位置，有且只能有一个。在视图布局页中，除 @section 部分外，其余的视图代码都将替换到这个位置。

2）RenderSection()：生成次要预留位置，可以有任意个，以 name 区分。常用的方法有 RenderSection（string name）和 RenderSection（string name, bool required）。在视图布局页中，@section name 对应的部分将替换到相应 name 的位置。required 表示是否必须要替换，默认情况下是 true，视图布局页中必须有相应的@section name 部分。

3）RenderPage()：生成其他视图的预留位置，可以有任意个。一般通过 RenderPage（string path）这个方法调用分部页（有关分部页的内容，将在 5.6 节中详细描述）。

下面通过一个例子来看一下布局页和视图布局页的使用情况。

【例 5.23】布局页和视图布局页的使用。

_MyLayoutPage.cshtml 中的代码（Shared 文件夹）如下。

```
1    <!DOCTYPE html>
2    <html>
3    <head>
4    <meta name="viewport" content="width=device-width" />
5    <title>@ViewBag.Title</title>
6    </head>
7    <body>
8    <div style="height:50px;background-color:black;">
9        @RenderSection("top",false)
10   </div>
11   <div>
12       @RenderBody()
13   </div>
14   <div style="height:50px;background-color:black">
15       @RenderPage("bottom.cshtml")
16   </div>
17   </body>
18   </html>
```

bottom.cshtml 中的代码（Shared 文件夹）如下。

```
<b style="color:white">这是页脚,通过页面调用</b>
```

MyView.cshtml 中的代码（Home 文件夹）如下。

```
1    @{
2        ViewBag.Title="MyView";
3        Layout="~/Views/Shared/_MyLayoutPage.cshtml";
4    }
5    @section top{
6        <b style="color:white">这是页头,也是副预留位置</b>
7    }
8    <h2>MyView</h2>
9    <div style="height:100px">这是主预留位置</div>
```

页面的运行结果如图 5.42 所示。

图 5.42　页面运行结果

查看源文件后，可得到如下的 HTML 代码。

```
1     <!DOCTYPE html>
2     <html>
3     <head>
4     <meta name="viewport" content="width=device-width" />
5     <title>MyView</title>
6     </head>
7     <body>
8     <div style="height:50px;background-color:black;">
9     <b style="color:white">这是页头,也是副预留位置</b>
10    </div>
11    <div>
12    <h2>MyView</h2>
13    <div style="height:100px">这是主预留位置</div>
14    </div>
15    <div style="height:50px;background-color:black">
16    <b style="color:white">这是页脚,通过页面调用</b>
17    </div>
18    </body>
19    </html>
```

5.5.4　布局页的嵌套

一般情况下，一系列的相关视图有一个布局页就够用了；但有时为了使用的方便，可以建立几个嵌套的布局页。如图 5.43 所示，布局页 A 是原始布局页，布局页 B 是布局页 A 的变型，即从"预留位置 1"中分离出一部分空间，形成"固定布局"，留下的空间变成"预留位置 1'"。在只有布局页 A 的情况下，若要形成布局页 B 的布局，就需要在视图布局页中编写"固定布局"的代码。当需使用布局 B 的视图布局页较多时，就会重复编写"固定布局"的代码;同时,在需要修改"固定布局"中的布局时,可能会造成遗漏。这时,使用布局页的嵌套就能解决这个问题:使用布局页 B "继承"布局页 A,同时对"预留位置1"进行拆分,取出一部分空间用于"固定布局"。(当然,也可以使用分部页来解决这个问题,有关分布的介绍将在 5.6 节中详细描述。)

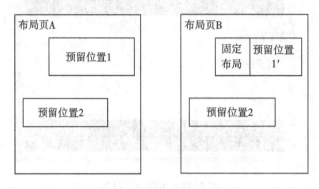

图 5.43 二级嵌套布局页示意图

在使用嵌套的布局页时,需要注意下面的两个问题。

1) 嵌套的布局页不能直接由系统生成,需要手动配置:在新建任意类型的视图后,清空原有的代码,编写@{ Layout = "上一级布局页的 URL";}和@ RenderBody()这两句代码。第一句表示"继承"的上一级布局页;第二句表示本级布局页中的主要预留位置。

2) 次要预留位置一般需要一并"继承",可以通过@ section name { @ RenderSection (name)} 来实现。

下面通过一个例子来看一下嵌套的布局页的使用情况。

【例 5.24】嵌套布局页的使用。

_MyLayoutPage. cshtml 中的代码同例 5.23。

_MyLayoutPage2. cshtml 中的代码(Shared 文件夹)如下。

```
1       @{
2           Layout = " ~/Views/Shared/_MyLayoutPage.cshtml";
3       }
4       @section top{
5           @RenderSection("top", false)
6       }
7       <div style = "height: 200px" >
8       <div style ="width: 100px; height: inherit; float: left; background-color: gray" >
9       <b style = "color: white" >这是第二个布局页中的布局 </b>
10      </div>
11      <div style = "float: left; padding-left: 10px" >@RenderBody() </div>
12      </div>
```

MyView. cshtml 中的代码(Home 文件夹)如下。

```
1       @{
2           ViewBag.Title = "MyView";
3           Layout = " ~/Views/Shared/_MyLayoutPage2.cshtml";
```

```
4        }
5        @section top{
6            <b style="color:white">这是页头,也是副预留位置</b>
7        }
8        <h2>MyView</h2>
9        <div style="height:100px">这是主预留位置</div>
```

页面的运行结果如图 5.44 所示。

图 5.44　页面运行结果

查看源文件后,可得到如下的 HTML 代码。

```
1    <!DOCTYPE html>
2    <html>
3    <head>
4        <meta name="viewport" content="width=device-width"/>
5        <title>MyView</title>
6    </head>
7    <body>
8        <div style="height:50px;background-color:black;">
9            <b style="color:white">这是页头,也是副预留位置</b>
10       </div>
11       <div>
12           <div style="height:200px">
13               <div style="width:100px;height:inherit;float:left;background-color:gray">
14                   <b style="color:white">这是第二个布局页中的布局</b>
15               </div>
16               <div style="float:left;padding-left:10px">
17                   <h2>MyView</h2>
18                   <div style="height:100px">这是主预留位置</div></div>
19           </div>
```

```
20      </div>
21      <div style = "height:50px;background-color:black">
22      <b style = "color:white">这是页脚,通过页面调用</b>
23      </div>
24      </body>
25      </html>
```

5.6 分部页

5.6.1 分部页的创建

在 MVC 4 中,分部页的创建过程与视图页的创建过程类似,可以通过以下 3 种方式来创建。

1. 通过控制器中的函数快速创建

Step01 ~ Step03:同 "通过控制器中的函数快速创建" 视图页过程(见 5.2.1 节)中的 Step01 ~ Step03。

Step04:勾选 "添加视图" 对话框中的 "创建为分部视图" 复选框,如图 5.45 所示。

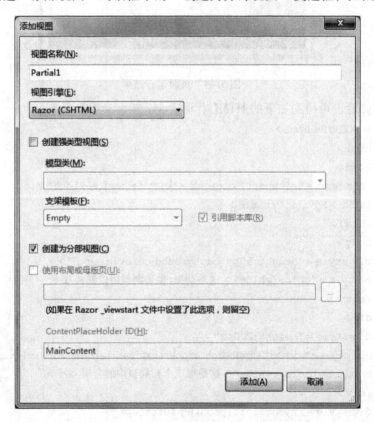

图 5.45 勾选 "创建为分部视图" 复选框

Step05:单击下方的 "添加" 按钮,即得到如图 5.46 所示的新创建的分部页页面。分部页在新建后一般都是空白页，无任何代码。

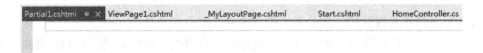

图 5.46　新创建的分部页页面

2. 通过"视图"创建

Step01 ~ Step03：同"通过'视图'创建"视图页过程（见 5.2.1 节）中的 Step01 ~ Step03。

Step04 ~ Step05：同"通过控制器中的函数快速创建"分部页过程中的 Step04 ~ Step05。

3. 通过"新建项"创建

Step01：同"通过'新建项'创建"视图页过程（见 5.2.1 节）中的 Step01。

Step02：单击子菜单中的第三项（即"新建项"），得到如图 5.47 所示的"添加新项"对话框。

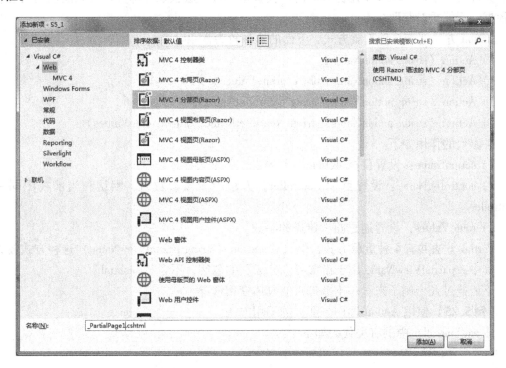

图 5.47　"添加新项"对话框

Step03：根据实际情况修改名称，本例名称修改为"_ PartialPage1. cshtml"。

Step04：单击下方的"添加"按钮，即得到如图 5.48 所示的新创建的分部页页面。

图 5.48　新创建的分部页页面

5.6.2 分部页介绍

分部页一般作为视图的一部分来使用,主要有以下 3 个功能。

第一,精简结构。当视图的构成较为复杂时,可以把一部分代码转移到分部页,通过调用分部页的方式来调用这些代码,使得主视图的结构比较清晰,见例 5.27。

第二,复用布局。当多个视图需要用到相同的页面布局时,可以把这些页面布局放到一个分部页中,这些视图可以通过调用分部页的方式复用页面布局。

第三,局部刷新。在视图运行中和用户交互时,可能需要通过局部刷新的方式提供一些信息,这些信息一般放在分部页中,提交给用户。

在 MVC 4 中,共有 4 种方法可以调用分部页:Action、RenderAction、Partial 和 RenderPartial。其中,由于@ Html.Action()等价于@ {Html.RenderAction();},@ Html.Partial()等价于@ {Html.RenderPartial();},因此本节只介绍 Action 和 Partial。

使用 Action 调用分部页时,会重新返回到控制器层获取有关数据等,之后才会从视图层获取分部页;使用 Partial 调用分部页时,会直接从视图层获取分部页,但只能使用当前视图的数据,不能重新获取。

Action 方法共有 6 种重载方式,下面的 4 种较为常用。

1) Action(string actionName)
2) Action(string actionName, object routeValues)
3) Action(string actionName, string controllerName)
4) Action(string actionName, string controllerName, object routeValues)

各参数的作用如下。

1) actionName:设置目标 Action。
2) controllerName:设置目标 Controller;方法中无该参数时,默认和当前视图同一个 Controller。
3) routeValues:设置通过路由传递的参数。

Partial 方法共有 4 种重载方式,但只有 Partial(string partialViewName)这种方式较为常用。其中,partialViewName 用于设置目标分部页的名字,不含 ".cshtml"。

下面通过几个例子来看一下分部页的具体使用情况。

【例 5.25】使用 Action 调用分部页的不用方法。

HomeController 中的有关代码如下。

```
1    public ActionResult S18()
2    {
3        return View();
4    }
5    public ActionResult Partial18(int id)
6    {
7        if (id! = null)
8        {
9            ViewBag.flag = true;
10           ViewBag.id = id;
```

```
11          }
12      else
13          ViewBag.flag = false;
14      return PartialView();
15  }
```

S18.cshtml 中的代码如下。

```
1   @{
2       Layout = null;
3   }
4   <!DOCTYPE html>
5   <html>
6   <head>
7       <meta name="viewport" content="width=device-width" />
8       <title>S18</title>
9   </head>
10  <body>
11      <div>
12          方法(1):@Html.Action("Partial18")
13          <br />
14          方法(2):@Html.Action("Partial18", new { id = 100 })
15          <br />
16          方法(3):@Html.Action("Partial18","Home")
17          <br />
18          方法(4):@Html.Action("Partial18","Home", new { id = 100 })
19      </div>
20  </body>
21  </html>
```

Partial18.cshtml 中的代码如下。

```
1   <div style="border:solid;">
2       这是分布视图.
3       <br />
4       @{
5           if(ViewBag.flag)
6           {
7               <span>传递过来的参数是 @ViewBag.id</span>
8           }
9       }
10  </div>
```

页面运行结果如图 5.49 所示。

【例 5.26】使用 Partial 调用分部页的方法。

HomeController 中的有关代码如下。

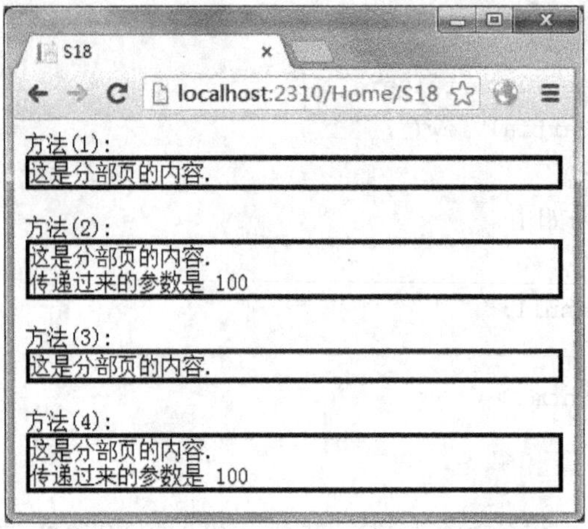

图 5.49　页面运行结果

```
1    public ActionResult S19()
2    {
3        //获取数据库连接
4        StudentEntities db = new StudentEntities();
5        //读取第一个学生的信息
6        Student s1 = db.Students.FirstOrDefault();
7        ViewData["S19"] = "这是通过 ViewData 传递的数据";
8        return View(s1);
9    }
```

S19.cshtml 中的代码如下。

```
1    @{
2        Layout = null;
3    }
4    <!DOCTYPE html>
5    <html>
6    <head>
7    <meta name = "viewport" content = "width = device-width" />
8    <title>S19</title>
9    </head>
10   <body>
11   <div>
12       这是主视图上的文字.
13           @Html.Partial("Partial19")
14   </div>
15   </body>
16   </html>
```

Partial19.cshtml 中的代码如下。

```
1    @model S5_1.Models.Student
2    <div style="border: solid;">
3    这是分部页的内容.
4    <br />
5        @ViewData["S19"]
6    <br />
7    学生信息：
8    <p>学号:@Model.Sno</p>
9    <p>姓名:@Model.Sname</p>
10   <p>性别:@Model.Ssex</p>
11   <p>年龄:@Model.Sage</p>
12   </div>
```

页面的运行结果如图 5.50 所示。

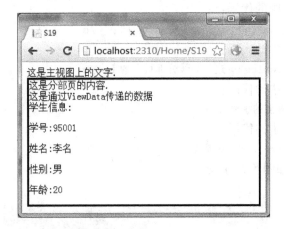

图 5.50　页面运行结果

【例 5.27】使用分部页精简结构。

主视图 MainView.cshtml 中的代码如下。

```
1    @{
2        Layout = null;
3    }
4    <!DOCTYPE html>
5    <html>
6    <head>
7    <meta name="viewport" content="width=device-width" />
8    <title>MainView</title>
9    </head>
10   <body>
11   <div style="height: 50px; background-color: black;">
12        @Html.Partial("MyPartial1")
13   </div>
14   <div style="height: 200px">
```

```
15        <div style="width: 100px; height: inherit; float: left; background-color: gray">
16              @Html.Partial("MyPartial2")
17        </div>
18        <div style="float: left; padding-left: 10px">
19              @Html.Partial("MyPartial3")
20        </div>
21    </div>
22    <div style="height: 50px; background-color: black">
23          @Html.Partial("MyPartial4")
24    </div>
25    </body>
26    </html>
```

4个分部页中的代码如下。

MyPartial1.cshtml 中的代码：
`<b style="color:white">这是1号分部页上的字</b>`

MyPartial2.cshtml 中的代码：
`<b style="color:white">这是2号分部页上的字</b>`

MyPartial3.cshtml 中的代码：
`<b>这是3号分部页上的字</b>`

MyPartial4.cshtml 中的代码：
`<b style="color:white">这是4号分部页上的字</b>`

页面的运行结果如图5.51所示。

图5.51 页面运行结果

5.7 向控制器层传递数据的方式

当视图在和用户进行交互以及在完成交互后，都需要把有关数据传递给控制器进行处理。在 MVC 4 中，有许多种传递方式，如路由、表单、JavaScript/JQuery、Session/Cookies 等。其中，路由方式一般通过 URL 传递，数据量不能过大；有关路由机制已在第 4 章中介绍。JavaScript/JQuery 方式一般用于局部刷新，将在第 8 章详细介绍。Seesion 和 Cookies 方式与从控制器层获取数据类似，已在 5.3.3 节详细介绍。因此，本节主要介绍表单方式。

表单对应的 HTML 标记是 < form > </form >。通常情况下，表单中的属性至少包含 action 属性（用于设置递交目标）和 method 属性（用于设置递交方式），如例 5.12 和例 5.13 生成的 HTML 代码。

当通过表单传递数据时，会先把 < form > 和 </form > 之间的数据整合到一个 FormCollection 中，然后传递给目标 Action，之后通过 FormCollection［name］获取相应的数据。当同一个表单中的数据有相同 name 时，在 FormCollection［name］中将会同时出现这几个数据，同时数据之间会用"，"隔开，可用 Split( ) 函数进行分割获取。

如果是通过 Post 方式传递，则需要在目标 Action 上方编写［HttpPost］，表明该 Action 用于 Post 方式。

下面通过一个例子来看一下如何利用表单方式传递数据。

【例 5.28】通过表单向控制器层传递数据。

HomeController 中的有关代码如下。

```
1    public ActionResult MyForm()
2    {
3        return View();
4    }
5    [HttpPost]
6    public ActionResult MyForm(FormCollection collection)
7    {
8        ViewBag.T1 = collection["T1"];
9        ViewBag.T2 = collection["T2"];
10       string T2 = collection["T2"];
11       string[] T2s = T2.Split(new char[] { ',' });
12       ViewBag.T2_1 = T2s[0];
13       ViewBag.T2_2 = T2s[1];
14       return View("MyForm2");
15   }
```

MyForm.cshtml 中的代码如下。

```
1    @{
2        Layout = null;
3    }
4    <!DOCTYPE html>
5    <html>
6    <head>
7        <meta name="viewport" content="width=device-width" />
8        <title>MyForm</title>
```

```
9       </head>
10      <body>
11          @{Html.BeginForm("MyForm");}
12   <p>@Html.TextBox("T1","这是文本输入框1中的数据", new { id = "S1" })</p>
13   <p>@Html.TextBox("T2","这是文本输入框2中的数据", new { id = "S2" })</p>
14   <p>@Html.TextBox("T2","这是文本输入框3中的数据", new { id = "S2" })</p>
15   <input type = "submit" value = "提交"/>
16          @{Html.EndForm();}
17      </body>
18   </html>
```

MyForm2.cshtml 中的代码如下。

```
1    @{
2           Layout = null;
3    }
4    <!DOCTYPE html>
5    <html>
6       <head>
7       <meta name = "viewport" content = "width = device-width" />
8       <title>MyForm2</title>
9       </head>
10      <body>
11   第一个输入框(name = "T1")传递过来的数据：
12       <br/>
13          @ViewBag.T1
14       <br/>
15   第二/三个输入框(name = "T2")传递过来的数据：
16       <br/>
17          @ViewBag.T2
18   拆分后的数据：
19       <br/>
20   第二个输入框：
21       <br/>
22          @ViewBag.T2_1
23       <br/>
24   第三个输入框：
25       <br/>
26          @ViewBag.T2_2
27      </body>
28   </html>
```

页面的初始运行结果如图5.52所示。表单包裹了当前页面的3个文本输入框，单击"提交"按钮后，表单将被提交，经控制器层处理后，页面运行结果如图5.53所示。

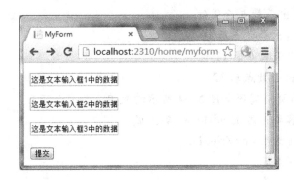

图 5.52　页面初始运行结果

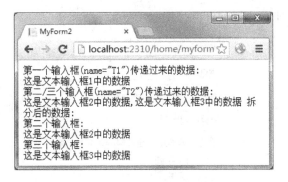

图 5.53　"提交"后的页面运行结果

※习　题

1. 新建一个视图页，实现下列功能：

1) 视图中，文字的初始状态如表 5.2 中的"初始状态"所示：第 1 行是"图形层数：X 层"，其中 X 随机生成，范围为 $1\sim10$，X 使用红色字体，加粗；第 $2\sim X+1$ 行由★组成，第 i 行有 $X+i-1$ 个★。

2) 单击第 1 行文字后，若 X 小于 10，则 X 变为 $X+1$，增加 1 行★；反之，X 变为 1，仅留下 1 行★。具体过程见表 5.2。

表 5.2　文字变化过程示意

初始状态	单击 1 次后	单击 2 次后	…	单击 9 次后	单击 10 次后
图形层数：2 层 ★ ★★	图形层数：3 层 ★ ★★ ★★★	图形层数：4 层 ★ ★★ ★★★ ★★★★	…	图形层数：1 层 ★	图形层数：2 层 ★ ★★

2. 在控制器层生成圆周率，分别通过下列方式将圆周率打印到同一个视图上。

1) ViewData，输出小数点后 8 位。

2）ViewBag，输出小数点后9位。

3）TempData，输出小数点后10位。

4）Session，输出小数点后11位。

5）Cookies，输出小数点后12位。

3. 新建一个视图页，实现如图5.54所示的表单，要求如下：

1）所有表单内容都通过HtmlHelper类生成。

2）表单提交地址为/Home/NewPeople。

3）密码不能明码显示。

4）性别只可选中一个。

5）年龄范围为18~30。

6）返回地址为/Home/index。

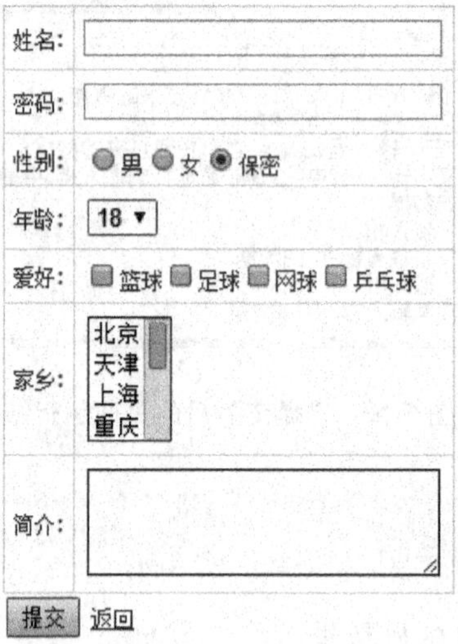

图 5.54　表单内容

4. 新建一套布局页和视图布局页，要求如下：

1）1个一级布局页。页面布局如图5.55所示，页眉处是系统名称；系统名称下方是若干个系统功能；系统功能下方是"预留位置"；页脚处是版权声明。（**注意**：图5.55~图5.60仅是示意图，系统名称、功能名称等需读者自行设定。）

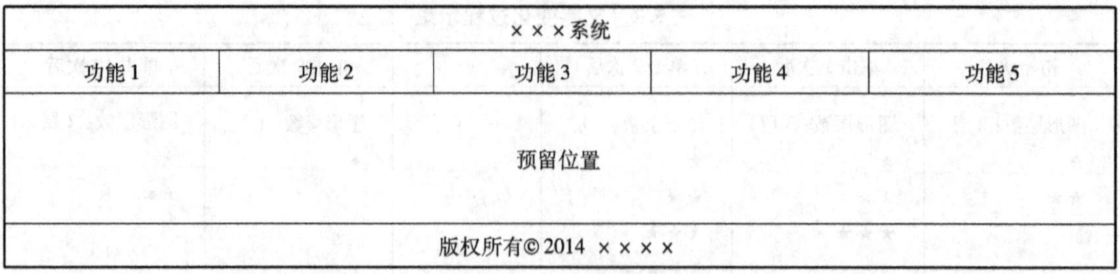

×××系统				
功能1	功能2	功能3	功能4	功能5
预留位置				
版权所有©2014 ××××				

图 5.55　一级布局页

2）两个二级布局页，"继承"自一级布局页。页面布局分别如图5.56和图5.57所示，将一级布局页中的"预留位置"分别取出左边的20%和右边的20%，用以放置若干个子功能。

×××系统				
功能1	功能2	功能3	功能4	功能5
子功能1	预留位置			
子功能2				
子功能3				
子功能4				
版权所有©2014 ××××				

图 5.56　二级布局页（左）

×××系统				
功能1	功能2	功能3	功能4	功能5
预留位置				子功能1
				子功能2
				子功能3
				子功能4
版权所有©2014 ××××				

图 5.57　二级布局页（右）

3）1个"继承"一级布局页的视图布局页，如图 5.58 所示。视图布局页中的内容使用习题 3 中的表单（读者亦可自行安排）。

×××系统				
功能1	功能2	功能3	功能4	功能5
表单内容				
版权所有©2014 ××××				

图 5.58　视图布局页（一级）

4）两个"继承"不同二级布局页的视图布局页，分别如图 5.59 和图 5.60 所示。视图布局页中的内容使用习题 3 中的表单（读者亦可自行安排）。

×××系统				
功能1	功能2	功能3	功能4	功能5
子功能1	表单内容			
子功能2				
子功能3				
子功能4				
版权所有©2014 ××××				

图 5.59　视图布局页（二级，左）

×××系统				
功能1	功能2	功能3	功能4	功能5
表单内容				子功能1
				子功能2
				子功能3
				子功能4
版权所有©2014 ××××				

图5.60 视图布局页（二级，右）

5. 将习题4中的"子功能"部分通过分部页实现。
6. 新建一个视图页，实现下列功能：
1）页面布局如图5.61所示。
2）表格中第2列数据由习题3中的表单提供，即在图5.54所示的表单中填入相关内容后，单击"提交"按钮，经控制器层处理，得到如图5.61所示的页面结果。
3）密码仅显示第1位和最后1位，中间部分用"*"代替。
4）"爱好"之间用"、"隔开。

姓名：	张三
密码：	0*****5
性别：	保密
年龄：	18
爱好：	篮球、足球、网球
家乡：	北京
简介：	我叫张三

图5.61 页面布局

※综合应用

本书将在第2、3和5章的综合应用中建立一个基本的拥有购物车功能的网站。其中，此处模块的重点主要在于View的编写。

本部分主要编写购物车的前台页面。前台部分主要包括Cart、Member、Order和Shared 4个模块。

Cart页面显示购物车中产品的名称、单价、数量、小计等信息。

```
1   @model List<MvcShopping.Models.Cart>
2   @{
3       var ajaxOption = new AjaxOptions() {
4           OnSuccess = "RemoveCartSuccess",
5           OnFailure = "RemoveCartFailure",
6           Confirm = "您确定要从购物车删除这个商品吗?",
7           HttpMethod = "Post"
8       };
9   }
10  @section scripts {
11      @Scripts.Render("~/bundles/jqueryval")
12      <script>
13      function RemoveCartSuccess() {
14          alert('移除购物车项目成功');
```

```
15                location.reload();
16            }
17        function RemoveCartFailure(xhr) {
18                alert('移除购物车项目失败 (HTTP 状态码：' + xhr.status + ')');
19            }
20    </script>
21    }
22    <h2>购物车清单</h2>
23    @using (Html.BeginForm("UpdateAmount", "Cart"))
24    {
25    <table>
26    <tr>
27    <th>产品名称</th>
28    <th>单价</th>
29    <th>数量</th>
30    <th>小计</th>
31    <th></th>
32    </tr>
33    @{int subTotal = 0; }
34    @for (int i = 0; i < Model.Count; i++)
35    {
36    //计算购买商品总价
37            subTotal += Model[i].Product.Price * Model[i].Amount;
38    //选择商品数量的选单只能选择 1 ～ 10
39    var ddlAmountList = newSelectList(Enumerable.Range(1, 10), Model[i].Amount);
40    …
41    <input type="button" value="完成订单"
42    onclick="location.href = '@Url.Action("Complete", "Order")';" />
43    </p>
44    }
```

Member 页面主要用于用户注册和登录，在第 3 章的综合应用中已经介绍过。

Order 页面主要用于订单的提交，页面显示需要提交的信息和备注。

```
1     @model MvcShopping.Models.OrderHeader
2     <h2>结账</h2>
3     @using (Html.BeginForm()) {
4     @Html.ValidationSummary(true)
5     <fieldset>
6     <legend>请输入派送信息与订单备注</legend>
7     <div class="editor-label">
8     @Html.LabelFor(model => model.ContactName)
9     </div>
10    <div class="editor-field">
11    @Html.EditorFor(model => model.ContactName)
```

```
12      @Html.ValidationMessageFor(model => model.ContactName)
13      </div>
14      <divclass = "editor-label">
15      @Html.LabelFor(model => model.ContactPhoneNo)
16      </div>
17      <divclass = "editor-field">
18      @Html.EditorFor(model => model.ContactPhoneNo)
19      @Html.ValidationMessageFor(model => model.ContactPhoneNo)
20      </div>
21      …
22      @sectionScripts {
23      @Scripts.Render("~/bundles/jqueryval")
24      }
```

Shared 页面是共享页面，主要用于退出和错误流操作。

退出页面代码如下。

```
1       <!DOCTYPEhtml>
2       <html>
3       <head>
4       <metacharset = "utf-8"/>
5       <metaname = "viewport"content = "width = device-width"/>
6       <title>@ViewBag.Title</title>
7       @Styles.Render("~/Content/css")
8       @Scripts.Render("~/bundles/modernizr")
9       </head>
10      <body>
11      <header>
12      <nav>
13      @Html.ActionLink("首页", "Index", "Home")
14              |@Html.ActionLink("购物车", "Index", "Cart")
15      @if (User.Identity.IsAuthenticated)
16              {
17      @: |@Html.ActionLink("登出", "Logout", "Member")
18              }
19      else
20              {
21      @: |@Html.ActionLink("登录", "Login", "Member")
22      @: |@Html.ActionLink("注册", "Register", "Member")
23              }
24      </nav>
25      </header>
26      @RenderBody()
27      @Scripts.Render("~/bundles/jquery")
28      @RenderSection("scripts", required: false)
```

```
29      </body>
30      </html>
```

错误信息页面代码如下。

```
1       @{
2           Layout = null;
3       }
4       <!DOCTYPEhtml>
5       <html>
6       <head>
7       <metahttp-equiv = "Content-Type"content = "text/html; charset = utf-8"/>
8       <metaname = "viewport"content = "width = device-width"/>
9       <title>错误</title>
10      </head>
11      <body>
12      <h2>
13      抱歉,处理您的要求时发生错误。
14      </h2>
15      </body>
16      </html>
```

第 6 章

ActionResult 类

本章将介绍如何利用 ASP. NET MVC 4 新增的区域（Area）机制构建较大的工程项目，以及如何将独立性较高的功能切割成一个 ASP. NET MVC 子网站，以降低网站之间的耦合性。同时，降低在多人同时开发一个项目时发生冲突的概率。

6.1 ActionResult 类概述

ActionResult 类是控制器层 Action 的返回值类型，它是一个抽象类，由它派生的实体类决定了该 Action 返回给视图层的具体类型（如常见的视图类型 ViewResult 等）。ActionResult 类及其派生类的类图如图 6.1 所示。各派生实体类的简要介绍见表 6.1。

表 6.1 ActionResult 派生实体类的简要介绍

序号	类名	返回内容	辅助方法
1	ViewResult	视图页、布局视图页	View
2	PartialViewResult	分部页	PartialView
3	ContentResult	文本内容	Content
4	EmptyResult	空白页	无，可以直接用 null
5	FileContentResult	通过 byte[ ]返回的文件	File
6	FileStreamResult	通过文件流返回的文件	File
7	FilePathResult	通过指定路径返回的文件	File
8	JavaScriptResult	JavaScript 对象	JavaScript
9	JsonResult	JSON 对象	Json
10	RedirectResult	重定向到指定的 URL	Redirect、RedirectPermanent
11	RedirectToRouteResult	重定向到指定的 Action	RedirectToAction、RedirectToActionPermanent、RedirectToRoute、RedirectToRoutePermanent
12	HttpUnauthorizedResult	登录页面	无
13	HttpNotFoundResult	HTTP 错误 404	HttpNotFound

注意：为了代码编写的方便，一般使用表 6.1 中的辅助方法来创建相应的 ActionResult 对象。例如，对于视图类型返回值，常用 return View( )来代替 return new ViewResult( )。

第 6 章 ActionResult 类

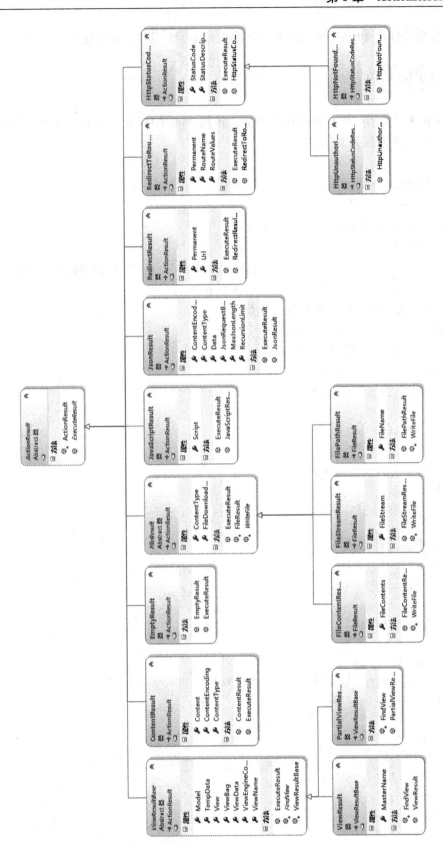

图6.1 ActionResult类及其派生类的类图

下面将分小节分别介绍各种 ActionResult 派生实体类的具体使用方法。

6.2 ViewResult

ViewResult 用于返回视图页和布局视图页，常用 View( ) 函数来实现。View( ) 函数有 8 种重载方式，下面 4 种较为常用。

1）View( )
2）View（string viewName）
3）View（object model）
4）View（string viewName，object model）

各参数的作用如下。

1）viewName：设置（布局）视图页的名称，用于在 Views 文件夹中搜索相应的（布局）视图页文件；方法 1) 中名称默认为所在 Action 的函数名。

2）model：设置传递给视图层的强类型数据。

下面通过例子来看一下 ViewResult 的使用情况。

【例 6.1】不使用强类型的 ViewResult。

HomeController 中的代码如下。

```
1    public ActionResult S1()
2    {
3        return View();
4    }
5    public ActionResult S2()
6    {
7        return View("S1");
8    }
```

S1.cshtml 中的代码如下。

```
1    @{
2        Layout = null;
3    }
4    <!DOCTYPE html>
5    <html>
6    <head>
7        <meta name="viewport" content="width=device-width" />
8        <title>S1</title>
9    </head>
10   <body>
11       <div>
12           这是页面 S1.
13       </div>
14   </body>
15   </html>
```

Action S1 和 S2 的页面的运行结果如图 6.2 所示。

图 6.2 页面运行结果

【例 6.2】使用强类型的 ViewResult。

HomeController 中的代码如下。

```
1       public ActionResult S3()
2       {
3           //获取第一个学生信息
4           Student s = db.Students.FirstOrDefault();
5           return View(s);
6       }
7       public ActionResult S4()
8       {
9           //获取第一个学生信息
10          Student s = db.Students.FirstOrDefault();
11          return View("S3",s);
12      }
```

S3.cshtml 中的代码如下。

```
1       @model S6_1.Models.Student
2       @{
3           Layout = null;
4       }
5       <!DOCTYPE html>
6       <html>
7       <head>
8           <meta name="viewport" content="width=device-width" />
9           <title>S3</title>
10      </head>
11      <body>
12      <div>
13          <p>学号:@Model.Sno</p>
14          <p>姓名:@Model.Sname</p>
15          <p>性别:@Model.Ssex</p>
16          <p>年龄:@Model.Sage</p>
17          <p>系所缩写:@Model.Sdept</p>
18      </div>
19      </body>
20      </html>
```

Action S3 和 S4 的页面的运行结果如图 6.3 所示。

图 6.3 页面运行结果

注意：只能搜索当前控制器所在的文件夹和 Shared 文件夹下的视图，可参考第 5 章关于视图的介绍。如果需要跨控制器复用视图时，则需要将视图放置在 Shared 文件夹下。

6.3 PartialViewResult

PartialViewResult 与 ViewResult 类似，但用于返回分部页，常用 PartialView()函数来实现。PartialView()函数有以下 4 种重载方式。

1）PartialView()

2）PartialView（string viewName）

3）PartialView（object model）

4）PartialView（string viewName, object model）

各参数的作用如下。

1）viewName：设置分部页的名称，用于在 Views 文件夹中搜索相应的分部页文件；方法 1）中的名称默认为所在 Action 的函数名。

2）model：设置传递给视图层的强类型数据。

由于 PartialView()和 View()的用法基本一致，因此本节不再重复举例介绍，可参考例 6.1、例 6.2 和例 5.25 ~ 例 5.27。

6.4 ContentResult

ContentResult 用于返回文本内容，常用 Content()函数来实现。Content()函数有以下 3 种重载方式。

1）Content（string content）

2）Content（string content, string contentType）

3）Content（string content, string contentType, System.Text.Encoding contentEncoding）

各参数的作用如下。

1）content：设置文本内容。

2) contentType：设置文本内容的 MIME 类型，如代表 CSS 样式的" text/css" 等。（有关 MIME 类型的详细介绍，读者可参考有关文献。）

3) contentEncoding：设置文本内容的编码方式，在中文系统中主要有 6 种，即 ASCII、BigEndianUnicode、Unicode、UTF32、UTF7 和 UTF8。

下面通过一个例子来看一下 ContentResult 的使用情况。

【例 6.3】ContentResult 的使用。

HomeController 中的代码如下。

```
1       public ActionResult S5()
2       {
3           return View();
4       }
5       public ActionResult S6()
6       {
7           return Content("这是文本内容!");
8       }
9       public ActionResult S7()
10      {
11          return Content("#S7{font-weight:bold; font-style: italic}", "text/css");
12      }
13      public ActionResult S8()
14      {
15          return Content("这是编码后的文本内容!", "text/html", System.Text.Encoding.UTF8);
16      }
```

S5.cshtml 中的代码如下。

```
1       @{
2           Layout = null;
3       }
4       <!DOCTYPE html>
5       <html>
6       <head>
7           <meta name="viewport" content="width=device-width" />
8           <title>S5</title>
9           <link type="text/css" rel="Stylesheet" href="@Url.Action("S7")" />
10      </head>
11      <body>
12          <div id="S7">
13              @Html.Action("S6")
14          </div>
15          @Html.Action("S8")
16      </body>
17      </html>
```

页面的运行结果如图6.4所示。

图6.4　页面运行结果

注意：要慎重选择编码方式，不合适的编码方式将导致页面乱码，如将S8中的编码方式改为ASCII时，页面将出现如图6.5所示的乱码。

图6.5　页面乱码

6.5　EmptyResult

EmptyResult用于返回空白页，常用null作为返回值，如下面的例6.4所示。

【例6.4】EmptyResult的使用。
HomeController中的代码如下。

```
1     public ActionResult S9()
2     {
3         return null;
4     }
```

页面的运行结果如图6.6所示。当查看网页源代码时，会发现没有任何代码，是名副其实的"空白页"。

图6.6　页面运行结果

6.6　FileContentResult、FileStreamResult和FilePathResult

FileContentResult、FileStreamResult和FilePathResult都用于返回文件，常用File()函数来实现，不同的重载方式决定了不同的类型。File()函数有以下6种重载方式。

1）FileContentResult File（byte [ ] fileContents，string contentType）

2）FilePathResult File（string fileName，string contentType）

3）FileStreamResult File（System. IO. Stream fileStream，string contentType）

4）FileContentResult File（byte [ ] fileContents，string contentType，string fileDownloadName）

5）FilePathResult File（string fileName，string contentType，string fileDownloadName）

6）FileStreamResult File（System. IO. Stream fileStream，string contentType，string fileDownloadName）

各参数的作用如下。

1）fileContents：设置比特流数据。

2）fileName：设置文件路径。

3）fileStream：设置流数据；Stream 是抽象类，派生出 MemoryStream、BufferedStream、FileStream 这 3 个实体类。

4）contentType：设置文件的 MIME 类型，如代表 PNG 图形文件的" image/png" 等。

5）fileDownloadName：设置文件的下载名称。

下面通过例子来分别看一下 FileContentResult、FilePathResult 和 FileStreamResult 的使用情况。

【例 6.5】FileContentResult 的使用。

HomeController 中的代码如下。

```
1    public ActionResult S10()
2    {
3        byte[] S10 = System.Text.Encoding.UTF8.GetBytes("这是通过比特流生成的文件!");
4        return File(S10, "text/plain","比特流文件.txt");
5    }
```

页面的运行结果如图 6.7 所示。保存后，打开下载的文件，能得到如图 6.8 所示的界面。

图 6.7　页面运行结果

图 6.8　下载文件的内容

【例 6.6】 FilePathResult 的使用。

HomeController 中的代码如下。

```
1       public ActionResult S11()
2       {
3           return File("/Content/空文件.xlsx", "application/vnd.ms-excel", "路径文件.xlsx");
4       }
```

页面的运行结果如图 6.9 所示。

图 6.9　页面运行结果

【例 6.7】 FileStreamResult 的使用。

HomeController 中的代码如下。

```
1       public ActionResult S12()
2       {
3           System.IO.FileStream S12 = new System.IO.FileStream(Server.MapPath
4       ("~/Content/空文件.xlsx"), System.IO.FileMode.OpenOrCreate);
5           return File(S12, "application/vnd.ms-excel", "流文件.xlsx");
6       }
```

页面的运行结果如图 6.10 所示。

第 6 章　ActionResult 类

图 6.10　页面运行结果

注意：一些特殊类型的文件，如图片类型，将在网页中直接显示。

6.7　JavaScriptResult

JavaScriptResult 用于返回 JavaScript 对象，常用函数 JavaScript（string script）来实现。下面通过一个例子来看一下 JavaScriptResult 的使用情况。

【例 6.8】JavaScriptResult 的使用。

HomeController 中的代码如下。

```
1       public ActionResult S13()
2       {
3           return View();
4       }
5       public ActionResult S14()
6       {
7           string S14 = "$(document).ready(function(){";
8           S14 = S14 + "alert('下面将通过 JQuery 添加网页内容！');";
9           S14 = S14 + "$('body').html('这是通过 JQuery 添加的网页内容！');";
10          S14 = S14 + "});";
11          return JavaScript(S14);
12      }
```

S13.cshtml 中的代码如下。

```
1   @{
2       Layout = null;
3   }
4   <!DOCTYPE html>
5   <html>
6   <head>
7       <meta name="viewport" content="width=device-width" />
8       <title>S13</title>
9       <script type="text/javascript" src="~/Scripts/jquery-1.7.1.js"></script>
10      <script type="text/javascript" src="@Url.Action("S14")"></script>
11  </head>
12  <body>
13      <div>
14      </div>
15  </body>
16  </html>
```

页面的初始运行结果如图 6.11 所示。单击"确定"按钮后，页面的运行结果如图 6.12 所示。

图 6.11 页面初始运行结果

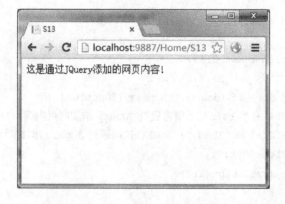

图 6.12 单击"确定"按钮后的页面

6.8　JsonResult

JsonResult 用于返回 JSON 对象，常用 Json()函数来实现。Json()函数有以下 6 种重载方式。

1）Json（object data）

2）Json（object data, JsonRequestBehavior behavior）

3）Json（object data, stringcontentType）

4）Json（object data, string contentType, JsonRequestBehavior behavior）

5）Json（object data, string contentType, System. Text. Encoding contentEncoding）

6）Json（object data, string contentType, System. Text. Encoding contentEncoding, JsonRequestBehavior behavior）

各参数的作用如下。

1）data：设置数据内容；一般的格式是｛"关键字 1":"值 1","关键字 2":"值 2",…, "关键字 n":"值 n"｝。有关 JSON 的详细介绍，读者可参考有关文献，本书不再赘述。在 MVC 4 中，强类型、结构体（类）、List<强类型>等具有一定结构的数据类型可以直接用作 data，不用转化成一般格式。

2）behavior：设置是否允许来自客户端的 http-get 请求，AllowGet 为允许，DenyGet 为不允许；出于安全的考虑，默认情况下 JSON 不接受 http-get 请求，即该参数为 DenyGet。

3）contentType：设置 MIME 类型，默认情况下是"application/json"；需要慎重修改 MIME 类型，不合适的类型将导致视图层得不到 JSON 对象。

4）contentEncoding：设置编码方式。

下面通过几个例子来看一下 JsonResult 的使用情况。

【例 6.9】JsonResult 的使用。

HomeController 中的代码如下。

```
1      public ActionResult S15()
2      {
3          return View();
4      }
5      public ActionResult S16()
6      {
7          //获取第一个课程的信息
8          Course s16 = db. Courses. FirstOrDefault();
9          return Json(s16);
10     }
11     public ActionResult S17()
12     {
13         //获取学生信息
14         List < Student > s17 = db. Students. ToList();
15         return Json(s17, "application/json", System. Text. Encoding. UTF8, Json Re-
16     questBehavior. AllowGet);
17     }
```

S15.cshtml 中的代码如下。

```
1      @{
2          Layout = null;
3      }
4      <!DOCTYPE html>
5      <html>
6      <head>
7          <meta name="viewport" content="width=device-width" />
8          <title>S15</title>
9          <script type="text/javascript" src="~/Scripts/jquery-1.7.1.js"></script>
10         <script type="text/javascript">
11             $(document).ready(function () {
12                 $.post("/Home/S16", function (data) {
13                     $("#S16").html("通过Json获取的课程信息");
14                     $("#S16").append("<p>名称:" + data.Cname + "</p>")
15                         .append("<p>学分:" + data.Ccredit + "</p>");
16                 });
17                 $.get("/Home/S17", function (data) {
18                     $("#S17").html("通过Json获取的学生信息");
19                     $("#S17").append("<table style='width:200px;' id='table'>");
20                     $("#table").append("<tr><td>学号</td><td>姓名</td><td>年龄</td><td>性别</td></tr>");
21                     for (var i = 0; i < data.length; i++) {
22                         $("#table").append("<tr id='tr" + i + "'>");
23                         $("#tr" + i).append("<td>" + data[i].Sno + "</td>")
24                             .append("<td>" + data[i].Sname + "</td>")
25                             .append("<td>" + data[i].Sage + "</td>")
26                             .append("<td>" + data[i].Ssex + "</td>")
27                             .append("</tr>");
28                     }
29                     $("#table").append("</table>");
30                 });
31             });
32         </script>
33     </head>
34     <body>
35         <div id="S16">
36         </div>
37         <div id="S17">
38         </div>
39     </body>
40     </html>
```

页面的运行结果如图 6.13 所示。

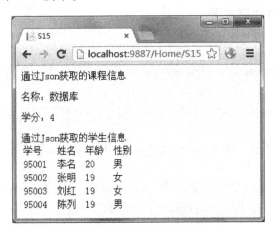

图 6.13　页面运行结果

注意：当 behavior 设置成 DenyGet，而视图层又通过 http-get 方式调用时，前台将得不到相应的 JSON 对象，将终止相应的 function( ) 函数。若在 S17 中将 behavior 设置成 DenyGet，则将得到如图 6.14 所示的运行结果。

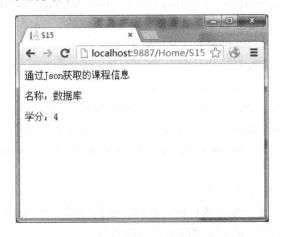

图 6.14　设置成 DenyGet 后的运行结果

6.9　RedirectResult

RedirectResult 用于重定向到指定的 URL，常用函数 Redirect（string url）和 RedirectPermanent（string url）来实现。前者用于 302 暂时性重定向，即 HTTP 302 状态，搜索引擎会抓取新的内容，但会保存旧的网址；后者用于 301 永久性重定向，即 HTTP 301 状态，搜索引擎在抓取新内容的同时，也会将旧的网址交换为重定向之后的网址（可参见百度百科相关介绍）。

下面通过一个例子来看一下 RedirectResult 的使用情况。

【例6.10】RedirectResult 的使用。

HomeController 中的代码如下。

```
1    public ActionResult S18()
2    {
3        return Redirect("/Home/S15");
4    }
5    public ActionResult S19()
6    {
7        return RedirectPermanent("/Home/S15");
8    }
```

页面运行后，在地址栏中无论输入"~/Home/S18"还是"~/Home/S19"（~表示前面的 localhost：9887 部分，因环境不同，端口号会发生变化，下同），经重定向后都将得到"~/Home/S15"。

6.10 RedirectToRouteResult

RedirectToRouteResult 用于重定向到指定的 Action，常用函数 RedirectToAction()、RedirectToRoute()、RedirectToActionPermanent()、RedirectToRoutePermanent()来实现。前两种函数用于 302 暂时性重定向，后两种函数用于 301 永久性重定向，但它们在使用方法上一一对应，因此本书只介绍前两种函数。

RedirectToAction()函数有 6 种重载方式，下面 4 种较为常用。

1) RedirectToAction (string actionName)
2) RedirectToAction (string actionName, object routeValues)
3) RedirectToAction (string actionName, string controllerName)
4) RedirectToAction (string actionName, string controllerName, object routeValues)

各参数的作用如下。

1) actionName：设置跳转的目标 Action。
2) controllerName：设置跳转的目标控制器，默认情况下是当前控制器。
3) routeValues：设置跳转时通过路由传递的参数。

RedirectToRoute()函数有 5 种重载方式，下面 3 种较为常用。

1) RedirectToRoute (object routeValues)
2) RedirectToRoute (string routeName)
3) RedirectToRoute (string routeName, object routeValues)

各参数的作用如下。

1) routeValues：设置跳转时通过路由传递的参数。
2) routeName：设置跳转时使用的路由的名称。

下面通过例子来看一下 RedirectToRouteResult 的使用情况。

【例6.11】RedirectToAction 的使用。

HomeController 中的代码如下。

```
1    public ActionResult S20()
2    {
3        return RedirectToAction("S15");
4    }
5    public ActionResult S21()
6    {
7        return RedirectToAction("S15", new { id = 1 });
8    }
9    public ActionResult S22()
10   {
11       return RedirectToAction("S15", "Home");
12   }
13   public ActionResult S23()
14   {
15       return RedirectToAction("S15", "Home", new { id = 1 });
16   }
```

页面运行后,在地址栏中输入"~/Home/S20"和"~/Home/S22",经重定向后将得到"~/Home/S15";输入"~/Home/S21"和"~/Home/S23",经重定向后将得到"~/Home/S15/1"。

【例 6.12】RedirectToRoute 的使用。

HomeController 中的代码如下。

```
1    public ActionResult S24()
2    {
3      return RedirectToRoute(new { controller = "Home", action = "S15", id = 1 });
4    }
5    public ActionResult S25()
6    {
7      return RedirectToRoute("Default2");
8    }
9    public ActionResult S26()
10   {
11     return RedirectToRoute("Default2", new { controller = "Home", action = "S15", id = 1 });
12   }
```

RouteConfig.cs 中新增的路由代码如下。

```
1    routes.MapRoute(
2        name: "Default2",
3        url: "{controller}/S1/{id}",
4        defaults: new { controller = "Home", id = UrlParameter.Optional }
5    );
```

页面运行后,在地址栏中输入"~/Home/S24",经重定向后将得到"~/Home/S15/1";输入"~/Home/S25",经重定向后将得到"~/Home/S1";输入"~/Home/S26",经重定向后将得到"~/Home/S1/1? action = S15"。

注意：因为在 Action S25 和 S26 中使用了新的路由"Default2"，而非默认路由"Default"，所以将按新的路由重定向，得到上述结果。

6.11　HttpUnauthorizedResult 和 HttpNotFoundResult

HttpUnauthorizedResult 用于返回 HTTP 错误 401，即未授权。因为返回未授权状态意义不大，所以一般会重定向登录页面。

HttpNotFoundResult 用于返回 HTTP 错误 404，即未找到。一般情况下，为了页面的美观，不会直接使用未找到页。

下面通过例子来分别看一下 HttpUnauthorizedResult 和 HttpNotFoundResult 的使用情况。

【例 6.13】HttpUnauthorizedResult 的使用。

HomeController 中的代码如下。

```
1    public ActionResult S27()
2    {
3        return new HttpUnauthorizedResult();
4    }
```

页面的运行结果如图 6.15 所示。

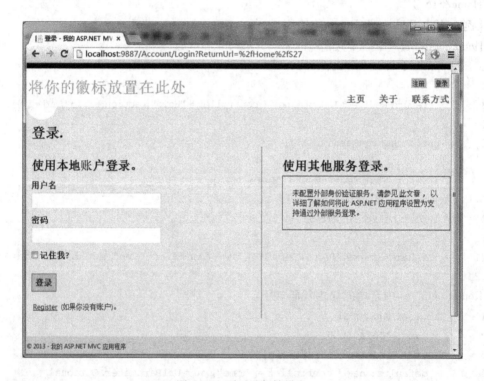

图 6.15　页面运行结果

【例 6.14】HttpNotFoundResult 的使用。

HomeController 中的代码如下。

```
1    public ActionResult S28()
2    {
3        return new HttpNotFoundResult();
4    }
```

页面的运行结果如图 6.16 所示。

图 6.16　页面运行结果

※习　题

1. 使用 PartialViewResult 实现例 6.1 和例 6.2。
2. 新建一个视图页，上面设置 3 个按钮，单击按钮后分别实现下列功能：

1）单击"按钮 1"后，下载一个文本文档，名字为"FileContent"，内容为"这是通过 FileContentResult 生成的文件！"。限定通过 FileContentResult 实现。

2）单击"按钮 2"后，下载一个 Word 文档（版本不限），名字为"FileStream"，内容为"这是通过 FileStreamResult 生成的文件！"。限定通过 FileStreamResult 实现。

3）单击"按钮 3"后，下载一个 PDF 文档（可事先准备好），名字为"FilePath"，内容为"这是通过 FilePathResult 生成的文件！"。限定通过 FilePathResult 实现。

3. 通过 JavaScriptResult 实现例 6.2。
4. 通过至少 4 种方式实现在地址栏中输入"~/Home/old"后转跳到"~/Home/new"。

※综合应用

说明在第 2、3 和 5 章综合应用部分完成的程序中使用了哪些 ActionResult 类,并尝试用相近的 ActionResult 代替上面使用的 ActionResult 类,注意,要达到相同的效果。

第 7 章

JavaScript 与 JQuery 技术

在 MVC 中，视图层原生的东西不是很多，视图层的制作主要是利用一些其他的网页制作技术。JavaScript 和 JQuery 在许多制作网页的方法中都会用到，但并不是 MVC 构架里特有的东西，相信许多读者并不陌生。本章讲解我们常用的前台技术：JavaScript 和 JQuery。我们从两种技术的基础开始，逐步深入到高级应用，并汇集一些高级应用的范例使之更容易使用。

7.1 JavaScript

JavaScript 是目前常用的前台页面技术。它直接在浏览器本地进行操作，基本上不需要向服务器发送指令，因而其在前台的展现、动态效果、前台控件等方面的应用较为强大，但是在数据交互和业务逻辑方面的应用较为薄弱。目前 JavaScript 主要用于实现前台布局和展现效果的应用。

7.1.1 JavaScript 简介

JavaScript 是一种网络脚本语言，它基于对象和驱动，用于完成网站的页面设计和实现表单验证等功能。在一般应用中，JavaScript 主要是作为一种实现 HTML 网页动态性和交互性等功能的工具来进行使用的。

使用中 JavaScript 的定义格式有两种，其中一种是直接写在页面当中。当 JavaScript 直接写在页面上时，它的位置是不固定的，既可以放在 <head> </head> 标签当中，又可以放在 <body> </body> 标签当中。但是一般情况下是放在 <head> </head> 标签中进行使用的。当在页面上直接编写 JavaScript 时，需要加入以下标签：

<script type = "text/javascript" > </script >

该标签声明目前使用的脚本语言是 JavaScript。代码和函数体则需要在两个标签之间进行编写。如下即为一段 JavaScript 程序，该代码表示会弹出引号中的文字。

```
1    < script type = "text/javascript" >
2       alert("这里是弹出的内容");
3    </script >
```

另外一种方式是将 JavaScript 写在 js 文件当中，然后在页面引用该 js 文件。如下是一个使用 js 文件的例子。

js 页面为 ShowAlert.js，代码如下。

```
1    window.onload = function()
2    {
3        Alert("This is a test!");
4    }
```

其中，window.onload 是初始化函数，主要作用是使随后的函数体在页面加载时就加载执行。这段代码的作用是在页面加载之初就显示"This is a test!"的字样。新建一个前台页面，添加这个 js 的应用，如下所示。

```
1    @{
2        ViewBag.Title = "View3";
3        Layout = "~/Views/Shared/_Layout.cshtml";
4    }
5    <script src="../../Scripts/ShowAlert.js" type="text/javascript"></script>
6    <h2>View3</h2>
```

其中，<script src="../../Scripts/ShowAlert.js" type="text/javascript"></script> 就是添加 js 文件的代码。这里使用的是相对地址，这个地址是依据文件的存储位置来定义的。"../"表示上一级目录。ShowAlert.js 文件存放的位置在前台页面文件的上上一级的 Scripts 文件夹下，故而文件应用格式为"../../Scripts/ShowAlert.js"。type 定义了文件的语言类型，这个 js 文件定义的是 JavaScript 语言，故而 type 的定义为"text/javascript"。页面的最终结果如图 7.1 所示。

图 7.1 ShowAlert.js 页面显示结果

这种方法和前一种方法在使用效果上是没有区别的，好处在于不用在同一个页面上添加过多的代码，使代码的清晰度较高。当 js 文件被页面应用后，浏览器执行的结果相当于把 JavaScript 代码写在页面当中。与直接写在页面上的方法不同的是，这种方法不需要添加 <script></script> 的标签，而是可以直接编写的。两种书写 JavaScript 的方法的区别见表 7.1。

表 7.1 两种书写 JavaScript 的方法的区别

	页面直接书写 JavaScript	将 JavaScript 写入 js 文件并引用
页面引用方式	<script type="text/javascript"></script>	<script src="../../Scripts/ShowAlert.js" type="text/javascript"></script>
特点	书写方便，查找容易	使界面整洁，代码工整
缺点	加大页面的篇幅，增加了页面代码读取的难度	无法使用 window.parent、top 等函数直接获得参数的值。如果多个 js 的前后顺序放置错误，则会对页面造成很大的干扰

总之，JavaScript 是一种轻量级的网络编程语言，它使用起来较为灵活，学习起来也非常方便，有任何一种程序设计语言基础的读者都可以很快掌握。

7.1.2 JavaScript 的语法

和任何一种编程语言一样，JavaScript 中的变量也是需要定义的。但是不同的是，JavaScript 的变量的定义较为简单，JavaScript 中的变量全部以"var"作为声明，不再区分 int、string、double、float 等变量类型。变量具体是什么类型是由给其赋予的值来决定的，如下面的代码所示。

```
1    <script type="text/javascript">
2        var string = "This type is string.";
3        var integer = 1;
4        var doubleNumber = 5.5;
5        var collum = [];
6    </script>
```

如上段代码所示，实际上 string 变量是字符串型的，integer 是整型变量，doubleNumber 是浮点型变量，collum 是数组。需要注意的是，给变量赋予一定的类型后，JavaScript 是允许改变变量为另一种类型的。

【例 7.1】在 JavaScript 中声明一个数组类型的变量 collum，然后将 collum 改变为整数，即可以赋予 12 并输出。

由于 JavaScript 对变量的规范程度较弱，因此，当一个变量定义好类型时，可以直接赋予其另外一种类型，代码如下所示。

```
1    window.onload = function() {
2        var collums = [];
3        collums = 12;
4        alert(collums);
5    }
```

运行结果如图 7.2 所示。

图 7.2 数组类型转变为整型的代码的运行结果

需要注意的是，由于 JavaScript 的这种特性的存在，使得代码的可读性和规范性变弱，因而在编写相关代码时需要格外注意。

JavaScript 语言的语法结构分为顺序结构、分支结构和循环结构。顺序结构较为容易理解，下面仅介绍分支结构和循环结构。

分支结构一般包含 if 语句或 switch 语句。if 语句使用较为频繁，其完整格式如下：

```
if (表达式) {语句集1}
else {语句集2}
```

当表达式为真时，执行"语句集1"的内容，当表达式为假时，执行"语句集2"的内容，其中 else 部分可以省略，当然也可以和其他语句进行嵌套使用。

switch 语句的格式如下：

```
switch(表达式)
{
case 情况1:
  语句集1
  break;
case 情况2:
  语句集2
  break;
  …
default:
  语句集n
}
```

当表达式结果为1时，执行"语句集1"的内容，当表达式结果为2时，执行"语句集2"的内容……以上全不成立时，执行"语句集n"的内容。

下面是一些分支语句的例子。

【例7.2】if 语句。

```
1      if (x < 60) {
2          alert("不及格");
3      }
4      else {
5          alert("及格");
6      }
```

和其他的语言相似，第1行和第4行是 if 语句的判定条件，如果 x < 60，则运行第2行命令，在屏幕上打印"不及格"，否则执行第5行，打印"及格"。

【例7.3】switch 语句。

```
1      switch (dayofaweek)
2      {
3      case 0:
4          x = " Sunday";
5          break;
6      case 1:
7          x = " Monday";
8          break;
9      case 2:
10         x = " Tuesday";
11         break;
```

```
12        case 3:
13                x = "Wednesday";
14                break;
15        case 4:
16                x = " Thursday";
17                break;
18        case 5:
19                x = "Friday";
20                break;
21        case 6:
22                x = "Saturday";
23                break;
24        }
```

switch 语句是选择式语句。

循环语句有 for 语句和 while 语句，其语法和 C 语言比较相似。其中 for 语句分为标准的 for 语句和 forin 语句，其中标准的 for 语句遵循如下格式：

```
for (初始化;条件;语句)
  {
     语句集
  }
```

forin 语句的格式如下：

```
for (变量 in 对象)
  {
     语句集
  }
```

for 语句的例子如下。

【例 7.4】 标准的 for 语句。

```
1        var sum=0;
2            for (var i=1; i<=10; i++) {
3                sum=sum + i;
4        }
```

【例 7.5】 forin 语句。

```
1        var person={fname:"Dave",lname:" Gates",age:25};
2        for (x in person)
3          {
4          all=all + person[x];
5          }
```

while 语句分为标准的 while 语句和 do/while 语句（JavaScript 中的 do/while 语句和一些语言的 do/until 语句比较相似）。标准的 while 语句的格式如下：

```
while (表达式)
{
语句集
```

}

当表达式为真时,执行"语句集"。

do/while 语句的格式如下:

do
{
语句集
}
while (表达式);

一直执行"语句集",直到表达式非真。

while 语句的例子如下。

【例 7.6】 while 语句。

```
1    var sum=0,i=1;
2    while (i<=10) {
3    sum=sum + i;
4    i=i + 1;
5    }
```

【例 7.7】 do/while 语句。

```
1    var sum=0,i=0;
2    do {
3        i=i + 1;
4        sum=sum + i;
5    }
6    while (i<=10);
```

需要注意的是,JavaScript 语言的执行顺序是严格按照顺序执行。当前一个语句有问题没有执行时,后面的语句是无法被执行的。并且由于 JavaScript 语言是一种前台语言,当语句出现错误时,在页面上除了没有达到预期的效果以外是不会有其他报错的现象的,因而,当页面没有达到预期效果时,需要注意这方面的内容。

7.1.3 JavaScript 函数

同其他语言一样,JavaScript 中也有函数,也就是我们经常提到的方法。JavaScript 中方法是用"function"进行定义的,如下所示。

```
function 函数名(参数1,参数2,…,参数n)
    {
        语句集
    }
```

需要注意的是,JavaScript 中函数名后面所带的参数是不用写 var 变量定义符的,这和其他语言在定义方法时不同。

【例 7.8】 定义一个名为 ShowMapLocation 的方法,其中有两个参数:Longitude、Latitude,用于输出显示 Longitude 和 Latitude 的值。

本例意在说明带有参数的 JavaScript 方法的使用方式,并最终以 alert 输出两个值,代码如下。

```
function ShowMapLocation(Longitude, Latitude){
    alert(Longitude + "," + Latitude);
}
```

另外，通过上面的例子可以看出，JavaScript 定义函数时是不需要定义函数的返回值的，但是这并不意味着 JavaScript 的函数没有函数值。JavaScript 依然使用 return 来进行值的返回，并且可以返回任何类型的值，如下所示。

```
function 函数名(参数1,参数2,...)
{
    语句集
    return 返回值;
}
```

当然，函数之间也可以进行嵌套调用，以及函数之间的相互调用。下面给出几个简单函数的例子。

【例7.9】无参数无返回值函数。

```
1    function alertans() {
2        alert("选项是Ⓐ,恭喜您答对了!");
3    }
```

【例7.10】有参数无返回值函数。

```
1    function alertans(a, b) {
2        if (a>b) alert("最大值是" + a);
3        else alert("最大值是" + b);
4    }
```

【例7.11】有参数有返回值函数。

```
1    function max(a, b) {
2        if (a>b) return a;
3        else return b;
4    }
5    var x = max(10,5);
```

【例7.12】函数嵌套。

```
1    function jiecheng(n) {
2        if (n==1) return 1;
3        else
4        return n * jiecheng(n-1);
5    }
6    var ans = jiecheng(10);
```

7.2 JQuery 简介

从本质上讲，JQuery 是 JavaScript 的一种库函数。相较于 JavaScript，这种动态库有兼容性好、交互性强、轻量级以及拥有很多成熟的插件等优点。并且相对于 JavaScript 而言，JQuery 本身的使用更为灵活，且对 CSS 支持度更高。基于以上种种原因，目前前台的应用当中，使用 JQuery 的较多。本节主要介绍一些 JQuery 的基础概念，并列举一些比较常用的方法供初学者参考。

在使用 JQuery 时，必须在页面上引入 JQuery 的库。首先需要下载 JQuery 库（最新的 JQuery 文件可以在 http：//jquery.com/中获得），放到相应的 Script 文件夹中，然后在 <head></head>之间加入如下引用：

```
<script src="../../Scripts/jquery-1.7.js"
type="text/javascript"></script>
```

从上面的应用可以看出，笔者所使用的 JQuery 库的版本是 1.7。JQuery 版本有很多，目前比较新的版本是 1.8.3 和 1.9.2 等版本。在 http://jquery.com/download/中可以下载到不同版本的 JQuery 库，在 http://www.w3school.com.cn/jquery/中可以查到 JQuery 的一些使用说明。

7.2.1 选择器

选择器是 JQuery 中很重要的一部分，主要用于指定需要操作的对象。相对于 JavaScript 的 document.getElementById("...")这种比较复杂的形式，选择器在获取对象上的操作极为简单。例如，需要获取"拥有 title 属性，并且值中包含 tested 的<aaa>元素"，只需要一条 JQuery 语句即可。可以说选择器是 JQuery 的精髓。

通俗地说，选择器就是"一个表示特殊语义的字符串"。只需要把选择器字符串传入，即可选择目标 Dom 对象，并且对象是以 JQuery 包装集的形式返回的。

下面我们来看一个例子。

【例 7.13】根据 ID 获取 JQuery 包装集。假设页面中有一个 text 元素，其 id 为 test，即为：<input type="text" id="test" value="获得文本内容"></input>。目前需要获得这个 input 集合和 input 的 value 值，代码形式如下所示。

1）var JQueryObj = $ ("#test")；
2）var JQueryObjValue = $ ("#test").val()；

JQuery 中最常用的便是根据 ID 获取 Dom 对象，上面的例子中代码 1）便是选取 ID 为 test 的 Dom 对象，并将选取的对象放入 JQuery 包装集，并且以 JQuery 包装集的形式返回。代码 2）是将包装集中的 value 属性的内容返回。"$"在 JQuery 中表示对 JQuery 库的引用。

选择器的类型有很多种分法，本书根据笔者在使用过程中的理解与体会，以功能为导向将选择器分为"选择"和"过滤"两大类。

"选择"类的选择器没有默认的选择范围，是在整个页面中进行选取操作的。而"过滤"类的选择器则以一定的条件为依据，对指定的内容进行筛选操作，当然"过滤"选择器也可以单独使用，表示从全部内容中筛选。对于"过滤"类选择器有如下约定：

```
$(":[title]");
```

等同于

```
$("* :[title]");
```

下面列举一些常用的选择器和过滤器，并给出对应的说明，见表 7.2～表 7.10。

表 7.2　基础选择器 Basics

名　称	说　明
#id	根据元素 ID 选择
element	根据元素的名称选择
.class	根据元素的 CSS 类选择
*	选择所有元素 $("*") 选择页面所有元素
selector1, selector2, ..., selectorN	可以将几个选择器用","分隔，然后拼成一个选择器字符串。会同时选中这几个选择器匹配的内容

表 7.3　层次选择器 Hierarchy

名　称	说　明
ancestor descendant	使用"form input"的形式选中 form 中的所有 input 元素，即 ancestor（祖先）为 form，descendant（子孙）为 input
parent > child	选择 parent 的直接子节点 child。child 必须包含在 parent 中，并且父类是 parent 元素
prev + next	prev 和 next 是两个同级别的元素。选中在 prev 元素后面的 next 元素
prev ~ siblings	选择 prev 后面的根据 siblings 过滤的元素。注：siblings 是过滤器

表 7.4　基本过滤器 Basic Filters

名　称	说　明
:first	匹配找到的第一个元素
:last	匹配找到的最后一个元素
:not(selector)	去除所有与给定选择器匹配的元素
:even	匹配所有索引值为偶数的元素，从 0 开始计数
:odd	匹配所有索引值为奇数的元素，从 0 开始计数
:eq(index)	匹配一个给定索引值的元素，index 从 0 开始
:gt(index)	匹配所有大于给定索引值的元素，index 从 0 开始
:lt(index)	选择结果集中索引小于 N 的 elements，index 从 0 开始
:header	选择所有 h1，h2，h3 一类的 header 标签
:animated	匹配所有正在执行动画效果的元素

表 7.5　内容过滤器 Content Filters

名　称	说　明
:contains(text)	匹配包含给定文本的元素
:empty	匹配所有不包含子元素或者文本的空元素
:has(selector)	匹配含有选择器所匹配的元素的元素
:parent	匹配含有子元素或者文本的元素

表 7.6 可见性过滤器 Visibility Filters

名　称	说　明
:hidden	匹配所有的不可见元素
:visible	匹配所有的可见元素

表 7.7 表单过滤器 Form Filters

名　称	说　明
:enabled	匹配所有可用元素
:disabled	匹配所有不可用元素
:checked	匹配所有被选中的元素（复选框、单选框等，不包括 select 中的 option）
:selected	匹配所有选中的 option 元素

表 7.8 属性过滤器 Attribute Filters

名　称	说　明
[attribute]	匹配包含给定属性的元素
[attribute = value]	匹配给定的属性是某个特定值的元素
[attribute != value]	匹配给定的属性是不包含某个特定值的元素
[attribute^= value]	匹配给定的属性是以某些值开始的元素
[attribute $ = value]	匹配给定的属性是以某些值结尾的元素
[attribute * = value]	匹配给定的属性是以包含某些值的元素
[attributeFilter1] [attributeFilter2] … [attributeFilter*N*]	复合属性选择器，需要在同时满足多个条件时使用

表 7.9 子元素过滤器 Child Filters

名　称	说　明
:nth-child（index/even/odd/equation）	匹配其父元素下的第 index 个子或奇偶元素：eq（index）'只匹配一个元素，而这个将为每一个父元素匹配子元素。：nth-child 从 1 开始的，而：eq()是从 0 算起的！可以使用： :nth-child(even) :nth-child(odd) :nth-child(3n) :nth-child(2) :nth-child(3n+1) :nth-child(3n+2)
:first-child	匹配第一个子元素：first'，只匹配一个元素，而此选择符将为每个父元素匹配一个子元素
:last-child	匹配最后一个子元素：last'，只匹配一个元素，而此选择符将为每个父元素匹配一个子元素
:only-child	如果某个元素是父元素中唯一的子元素，那么将会被匹配；如果父元素中含有其他元素，那么将不会被匹配

表 7.10　表单选择器 Forms

名　　称	说　　明
:input	匹配所有 input、textarea、select 和 button 元素
:text	匹配所有的文本框
:password	匹配所有密码框
:radio	匹配所有单选按钮
:checkbox	匹配所有复选框
:submit	匹配所有提交按钮
:image	匹配所有图像域
:reset	匹配所有重置按钮
:button	匹配所有按钮
:file	匹配所有文件域

7.2.2　JQuery 中的文件对象模型与方法

文件对象模型（Document Object Model，DOM）是指 HTML 中的元素的集合。7.2.1 节中的选择器选择的元素就是 Dom 元素。在上面已经提到过，选择器是 JQuery 的核心内容，在选择出元素后，需要进行相应的加工、处理，才能实现我们需要的功能。本节主要以常见的 DIV 元素为代表，列举一些常用的操作，并在后面附上大部分方法的说明。

【例 7.14】 改变 DIV（ID 为 a1，以下不再说明）的宽度和高度。

方法 1：

```
$("#a1").width(200);  //将其宽度变为 200px
$("#a1").height(100); //将其高度变为 100px
```

方法 2：

```
$("#a1").css({"width":"200px"});  //将其宽度变为 200px
$("#a1").css({"height":"100px"}); //将其高度变为 100px
```

【例 7.15】 获取 DIV 的宽度和高度。

```
Var height = $("#a1").width(); //获取 DIV 的宽度
$("#a1").height();             //获取 DIV 的高度
```

【例 7.16】 显示、隐藏 DIV。

```
$("#a1").show(); //显示 DIV
$("#a1").hide(); //隐藏 DIV
```

【例 7.17】 删除 10 个 DIV，ID 分别为 a1，a2，a3…a10。

```
For (var i=1;i<=10;i++)
$("#a"+i).remove(); //删除 10 个 DIV
```

【例 7.18】 改变 DIV 的 z-index 属性。

```
$("#a1").css({"position":"absolute",
"z-index":"2"});
```

注意： 在使用 z-index 属性时，需要先将 position 属性设置为"absolute"，否则这个属性将不起作用。

JQuery 中有一个非常有用的函数：attr。传入参数不同，该函数的工作方式也有所差异。其中一种使用 attr 提供返回值的工作方式的语法为：

Var xx = $(select).attr(attribute);

另一种使用 attr 设置属性域值的语法为：

$(select).attr(attribute,value);

下面通过示例对 attr 函数进行简要说明。

【例 7.19】判断 checkbox（ID 为 selectallcheck）是否被选中。

```
If($("#selectallcheck").attr("checked")==true)
    alert("yes");//被选中
else
alert("no");//未被选中
```

【例 7.20】将图片（ID 为 img1）的 alt 属性设为 1。

$("#img1").attr("alt","1"); //设置 alt 属性

除了 attr 方法以外，JQuery 的常用方法见表 7.11～表 7.19。

表 7.11　内部插入方法

方　　法	说　　明
append( )	在每个匹配元素的结尾插入内容
appendTo( )	匹配元素集合中的每个元素将被插入指定的元素末尾
prepend( )	在每个匹配元素的开头插入内容
prependTo( )	匹配元素集合中的每个元素将被插入指定的元素开头
html( )	有 html( )、html(htmlString) 和 (function(index,html)) 三种： 无参数的方法为获取匹配元素的内容； htmlString 表示要为每个匹配元素设置要替换的内容； function(index,html) 是一个函数，它返回一个 HTML 字符串，并将其替换到每个匹配元素的内容中
text( )	有 text( )、text(textString) 和 (function(index,text)) 三种： 无参数的方法为获取匹配元素的内容； textString 表示要为每个匹配元素设置要替换的内容； function(index,text) 是一个函数，它返回一个 text 字符串，并将其替换到每个匹配元素的内容中

表 7.12　外部插入方法

方　　法	说　　明
after( )	在每个匹配元素后插入内容
insertAfter( )	每个匹配元素都将插入到由参数指定的元素之后
before( )	在每个匹配元素前插入内容
insertBefore( )	每个匹配元素都将插入到由参数指定的元素之前

表 7.13　删除元素

方　　法	说　　明
remove( )	彻底删除所匹配元素
detach( )	在删除匹配的元素时，保留了其所有相关的 JQuery 信息
empty( )	删除匹配元素的所有后代节点

表 7.14 替换和复制元素

方法	说明
replaceWith( )	替换匹配的内容
replaceAll( )	替换目标元素
clone( )	复制元素 clone([true/false]) 当参数为 true 时,表示同时复制元素包含的事件。如果没有参数,默认为 false

表 7.15 包装元素

方法	说明
wrap( )	包装匹配元素
unwrap( )	wrap( )的逆过程,解除元素的包装
wrapAll( )	将所有匹配的元素进行一次包装
wrapInner( )	包装匹配元素的内容

表 7.16 设置和获取 DOM 属性

方法	说明
attr( )	attr(attributeName)为获取 DOM 元素的属性,获取匹配元素集合中的第一个元素的名为 attributeName 的属性值 attr(attributeName,value)为设置 DOM 元素的属性,设置名为 attributeName 的属性值为 value
removeAttr( )	删除属性
val( )	val( )表示获取匹配元素的 value 属性 val(newval)表示设定匹配元素的 value 值为 newval

表 7.17 CSS 相关的方法

方法	说明
css( )	css(propertyName)为获取名为 propertyName 的 CSS 属性值 css(propertyName,value)为设置名为 propertyName 的 CSS 属性值为 value
addClass( )	在匹配的元素中添加指定的 CSS 类
hasClass( )	判断匹配的元素中是否包含一个 CSS 类,如果包含,则返回 true,反之返回 false
removeClass( )	删除匹配元素中指定的 CSS 类
toggleClass( )	此方法是 addClass( )和 removeClass( )的整合方法,语法为 toggle(className[,switch])。当 switch = true 时,相当于 addClass( );当 switch = false 时,相当于 removeClass( )

表 7.18 设置和获取元素的位置

方法	说明
offset( )	offset( )为获取匹配元素相对于文档的当前坐标,返回一个含 top 和 left 的对象 offset(coordinates)为设置匹配元素相对于文档的当前坐标
position( )	获取匹配元素集合中的第一个元素相对于父元素的当前坐标,返回一个含 top 和 left 的对象
scrollLeft( )	scrollLeft( )为获取匹配元素的水平滚动条的位置 scrollLeft(value)为设置获取匹配元素的水平滚动条的位置
scrollTop( )	scrollTop( )为获取匹配元素的垂直滚动条的位置 scrollTop(value)为设置获取匹配元素的垂直滚动条的位置

表 7.19 设置和获取元素的大小

方法	说明
height( )	height( )为获取匹配元素的高度 height(value)为设置匹配元素的高度为 value
width( )	width( )为获取匹配元素的宽度 width(value)为设置匹配元素的宽度为 value
innerHeight( )	获取元素本身的高度+填充层(padding)的高度的值
innerWidth( )	获取元素本身的宽度+填充层(padding)的宽度的值
outerHeight( )	outerHeight([true/false]) true：获取元素本身的高度+填充层(padding)的高度+边框(border)的高度+边距(margin)高度的值 false：获取元素本身的高度+填充层(padding)的高度+边框(border)的高度的值
outerWidth( )	outerWidth([true/false]) true：获取元素本身的宽度+填充层(padding)的宽度+边框(border)的宽度+边距(margin)宽度的值 false：获取元素本身的宽度+填充层(padding)的宽度+边框(border)的宽度的值

7.2.3 事件处理

事件(event)是即时交互的触发者，JQuery 中的事件和 JavaScript 中的事件类似，只是在书写形式上去掉了 JavaScript 中的 "on" 的字样，如 JavaScript 中的 onclick 事件在 JQuery 中为 click 事件。对于 JQuery 中的选择器，Dom 元素的处理，或者这里说的事件，都是没有必要完全记下来的，只需要记住几种比较常用的方法，对于其他的有所了解即可，有需要的时候再到 JQuery 官方手册中进行查找即可。

下面将列举一些常用的 JQuery 事件。

1) click( )事件：鼠标单击事件。用法如下：

```
1    $('#btn1').click(function(){
2    $('#a1').hide();});
```

该例的含义是：当触发 ID 为 btn1 的元素（不一定是按钮，可能是文字等元素）的单击事件后，ID 为 a1 的 DIV 将会被隐藏。整个过程涉及 click、mousedown、mouseup 事件，如果三个事件都进行了定义，其顺序是 mousedown→mouseup→click。

2) bind( )事件：表示当某个对象失去焦点时所触发的事件。

3) change( )事件：表示当符合定义的对象失去焦点并且其值发生改变时触发的事件。

4) dblclick( )事件：表示鼠标双击某个对象时所触发的事件。

5) error( )事件：表示当 error 发生时所触发的事件。用法如下：

$(window).error(function(){return true;});

上述代码表示当 window 有 error 时（如网页有错误），显示错误信息。

6) focus( )事件：表示当对象获得焦点时所触发的事件。

7) hover(over,out)事件：表示当鼠标移动到相应对象上或离开相应对象时所触发的事件，必须同时定义 over（指向）以及 out（离开）这两项内容。

8）keydown( )事件：定义键盘上某个键按下时所触发的事件。用法如下：

```
1    $(document).keydown(function(e){
2    $('#a1').hide()});
```

上述代码表示当我们按下键盘上的按键时会隐藏 ID 为 a1 的 DIV。

另外，mousedown( )、mouseup( )、keypress( )的功能与 keydown( )的功能类似，这里不再赘述。

7.3 JavaScript 与 JQuery 应用实例

本节开始介绍 JavaScript 和 JQuery 的应用实例。

7.3.1 iPhone 界面制作

本小节将介绍一个 JQuery 的高级应用——使用 JQuery 实现 iPhone 的界面效果。这里需要使用 JQuery 的 Hover 函数。在本小节中，我们将在 HTML 网页中添加一个 3 行 3 列，可进行左右滑动的 iPhone 界面。具体效果如图 7.3 所示。

a）

b）

图 7.3　利用 JQuery 实现 iPhone 界面的效果图

在这里，需要使用 HTML 页面上元素属性的遮照效果、CSS 样式、鼠标单击移动效果、页面图标排序效果、JQuery 的 hover 事件等。在 JQuery 事件当中，定义显示区域的宽度、高度，图标和文字的绑定，图标和单击链接的绑定，鼠标操作的相应时间等事件。具体代码如下所示。

```
1    <!DOCTYPE html PUBLIC "-//W3C//DTD XHTML 1.0 Transitional//EN"
2    "http://www.w3.org/TR/xhtml1/DTD/xhtml1-transitional.dtd">
3    <html xmlns="http://www.w3.org/1999/xhtml">
4        <head>
5        <meta http-equiv="Content-Type" content="text/html;charset=utf-8" />
6        <title>JQuery 苹果 iOS 手机主屏幕触摸效果</title>
7        <script src="http://html5shiv.googlecode.com/svn/trunk/html5.js">
8        </script>
9        </head>
10       <body>
11       <style type="text/css">
12       *{margin:0;padding:0;}
13       html{background-color:#161616;}
14       body{color:#fff;min-height:600px;font-family:Arial;}
15       a,a:visited {text-decoration:none;outline:none;color:#54a6de;}
16       a:hover{text-decoration:underline;}
17       /* ------------------------------ 主屏幕 ------------------------------ */
18       #homeScreen{width:810px;height:770px;padding:1px;margin:0 auto 30px;
19       background:url('images/background.jpg') no-repeat left bottom;}
20       /* mask 只显示在一个屏幕上的事件。使用溢出:隐藏 */
21       #mask{width:332px;height:380px;position:relative;overflow:hidden;mar-
22       gin:180px auto 0;}
23       #allScreens{height:100%;top:0;left:0;position:absolute;cursor:move;}
24       .screen{width:332px;float:left;}
25       #dock .dockicon,.screen .icon{float:left;width:60px;height:60px;back-
26       ground-repeat:no-repeat;margin:25px;position:relative;}
27       /* 下面的图标显示的标题属性:*/
28       .screen .icon:after{bottom:-25px;color:white;content:attr(title);font-
29       size:12px;height:20px;left:-20px;overflow:hidden;position:absolute;text-
30       align:center;white-space:nowrap;width:100px;text-shadow:0 0 3px #222;}#dock
31       {height:70px;margin:60px auto;width:332px;}
32       /* dock 栏上的图标得到下面这些细微的阴影 */
33       #dock .dockicon:after{border-radius:50px/10px;bottom:7px;box-shad
34       ow:0 5px 2px #000000;content:"";height:1px;position:absolute;
35       width:58px;}
36       #indicators{text-align:center;list-style:none;}
37       #indicators li {border-radius:50%;display:inline-block;margin:7px;
38       width:6px;height:6px;background-color:white;opacity:0.6;}
39       #indicators li.active{opacity:1;background-color:#00A2D6;box-shad ow:0 0
40       3px #00A2D6, 0 0 1px #51CFF9 inset;}
41       </style>
```

```
42        <div id="homeScreen">
43            <div id="mask">
44                <div id="allScreens"></div>
45            </div>
46            <ul id="indicators"></ul>
47            <div id="dock"></div>
48        </div>
49        <script type="text/javascript" src="js/jquery-1.4.2.min.js"></script>
50        <script type="text/javascript" src="js/touchable.js"></script>
51        <script type="text/javascript" src="js/coffee-script.js"></script>
52        <script type="text/coffeescript">
53            # The Icon class.
54            class Icon
55                # The constructor. The -> arrow signifies
56                # a function definition.
57                constructor: (@id, @title) ->
58                    # @ is synonymous for "this". The id and title parameters
59                    # of the construtor are automatically added as this.id and this.title
60                    # @markup holds the HTML of the icon. It is
61                    # transformed to this.markup behind the scenes.
62                    @markup = "<div class='icon' style='background-image:url
63 (images/icons/#{@id}.png)'
64                        title='#{@title}'></div>"
65            # The DockIcon class inherits from Icon
66            class DockIcon extends Icon
67                constructor: (id, title) ->
68                    # This calls the constructor if Icon
69                    super(id, title)
70                    # Changing the class name of the generated HTML
71                    @markup = @markup.replace("class='icon'","class='dockicon'")
72            # The Screen Class
73            class Screen
74                # Function arguments can have default parameters
75                constructor: (icons=[]) ->
76                    @icons = icons
77                attachIcons: (icons=[]) ->
78                    Array.prototype.push.apply(@icons, icons)
79                generate: ->
80                    markup = []
81                    # Looping through the @icons array
82                    markup.push(icon.markup) for icon in @icons
83                    # The last line of every function is implicitly returned
84                    "<div class='screen'>#{markup.join('')}</div>"
```

```
85      class Stage
86          # The width of our "device" screen. This is
87          # basically the width of the #mask div.
88          screenWidth: 332
89          constructor: (icons) - >
90              @currentScreen = 0
91              @screens = []
92              # Calculating the number of screens
93              # necessary to display all the icons
94              num = Math.ceil(icons.length / 9)
95              i = 0
96          while num --
97              # we pass a slice of the icons array
98              s = new Screen(icons[i...i + 9])
99              # adding the screen to the local screens array
100             @screens.push(s)
101             i + = 9
102         # This method populates the passed element with HTML
103         addScreensTo: (element) - >
104             @element = $ (element)
105             @element.width(@screens.length* @screenWidth)
106             for screen in @screens
107                 @element.append(screen.generate())
108         addIndicatorsTo: (elem) - >
109             # This method creates the small
110             # circular indicatiors
111             @ul = $ (elem)
112             for screen in @screens
113                 @ul.append(' < li > ')
114             @ul.find('li:first').addClass('active');
115         goTo: (screenNum) - >
116             # This method animates the allScreen div in
117             # order to expose the needed screen in #mask
118             if @element.is(':animated')
119                 return false
120             # if this is the first or last screen,
121             # run the end of scroll animation
122             if @currentScreen == screenNum
123                 # Parallel assignment:
124                 [from, to] = [' + =15',' - =15']
125                 if @currentScreen ! =0
126                     [from, to] = [to, from]
127                 @element.animate( { marginLeft : from }, 150 )
```

```
128                    .animate( { marginLeft : to }, 150 )
129            else
130                # If everything is ok, animate the transition between the screens.
131                # The fat arrow = > preserves the context of "this"
132            @element.animate ( { marginLeft:-screenNum * @screenWidth }, = > @cur-
133     rentScreen = screenNum )
134            @ul.find('li').removeClass('active').eq(screenNum).addClass('active');
135        next: - >
136            toShow = @currentScreen + 1
137            # If there is no next screen, show
138            # the current one
139            if toShow == @screens.length
140                toShow = @screens.length - 1
141            @goTo(toShow)
142        previous: - >
143            toShow = @currentScreen - 1
144            # If there is no previous screen,
145            # show the current one
146            if toShow == -1
147                toShow = 0
148            @goTo(toShow)
149     # This is equivalent to $('document').ready(function(){}):
150     $ - >
151        allIcons = [
152            new Icon('Photos', 'Photo Gallery'), new Icon('Maps', 'Google Maps')
153            new Icon('Chuzzle', 'Chuzzle'), new Icon('Safari', 'Safari')
154            new Icon('Weather', 'Weather'), new Icon('nes', 'NES Emulator')
155            new Icon('Calendar', 'Calendar'), new Icon('Clock', 'Clock')
156            new Icon('BossPrefs', 'Boss Prefs'), new Icon('Chess', 'Chess')
157            new Icon('Mail', 'Mail'), new Icon('Phone', 'Phone')
158            new Icon('SMS', 'SMS Center'), new Icon('Camera', 'Camera')
159            new Icon('iPod', 'iPod'), new Icon('Calculator', 'Calculator')
160            new Icon('Music', 'Music'), new Icon('Poof', 'Poof')
161            new Icon('Settings', 'Settings'), new Icon('YouTube', 'Youtube')
162            new Icon('psx4all', 'PSx4All'), new Icon('VideoRecorder', 'Re-
163     cord Video')
164            new Icon('Installer', 'Installer'), new Icon('Notes', 'Notes')
165            new Icon('RagingThunder', 'RagingThunder'), new Icon('Stocks', 'Stocks')
166            new Icon('genesis4iphone', 'Genesis'), new Icon('snes4iphone', '
167     SNES Emulator')
168            new Icon('Calendar', 'Calendar'), new Icon('Clock', 'Clock')
169            new Icon('Photos', 'Photo Gallery'), new Icon('Maps', 'Google Maps')
170        ]
```

```
171            dockIcons = [
172                new DockIcon('Camera', 'Camera')
173                new DockIcon('iPod', 'iPod')
174                new DockIcon('Calculator', 'Calculator')
175            ]
176            allScreens = $('#allScreens')
177            # Using the Touchable plugin to listen for
178            # touch based events:
179            allScreens.Touchable();
180            # Creating a new stage object
181            stage = new Stage(allIcons)
182            stage.addScreensTo(allScreens)
183            stage.addIndicatorsTo('#indicators')
184            # Listening for the touchablemove event.
185            # Notice the callback function
186            allScreens.bind 'touchablemove', (e,touch) ->
187                stage.next() if touch.currentDelta.x < -5
188                stage.previous() if touch.currentDelta.x > 5
189            # Adding the dock icons:
190            dock = $('#dock')
191            for icon in dockIcons
192                dock.append(icon.markup)
193        </script>
194    </body>
195  </html>
```

在编写 JQuery 的 hover 事件时，需要注意每行的代码前后的空格数量，在 hover 的编写时，代码的对齐关系不能随意改动，所以不能随意删除前后的空格。当空格的数量不对时，代码结果也可能会出现问题。

例 7.22 修改以上代码，在页面中图标总数不变的情况下改为每页 2 行，每行 2 个图标，共每页 4 个图标的样式。

7.3.2 使用 JQuery 给 table 动态添加、删除行

页面上一种比较难的操作是对已经生成好的 table 的操作。对于生成好的 table，我们在页面上是无法直接动态添加行的。在这里，我们提供一种使用 JQuery 来动态添加行的方法，即使用 jquery-1.4.2_min.js 的 JQuery 库来支持操作的完成。相关代码如下所示。

```
1    <html>
2      <head>
3        <title>
4        </title>
5        <script src="js/jquery-1.4.2_min.js" type="text/javascript"></script>
6        <script type="text/javascript" language="javascript">
7            var row_count = 0;
```

```
 8              function addNew()
 9              {
10                  var table1 = $('#table1');
11                  var firstTr = table1.find('tbody > tr:first');
12                  var row = $("<tr></tr>");
13                  var td = $("<td></td>");
14                  td.append($("<input type='checkbox' name='count' value='New'><b>CheckBox" + row_count + "</b>"));
15
16                  row.append(td);
17                  table1.append(row);
18                  row_count++;
19              }
20              function del()
21              {
22                  var checked = $("input[type='checkbox'][name='count']");
23
24                  $(checked).each(function(){
25                      if($(this).attr("checked") == true)
26                      //注意:此处不能用$(this).attr("checked") == 'true'来判断
27                      {
28                          $(this).parent().parent().remove();
29                      }
30                  });
31              }
32          </script>
33      </head>
34      <body>
35          <input type="button" value="Add" onclick="addNew();">
36          <input type="button" value="Delete" onclick="del();">
37          <div id="rightcontent">
38              <table id="table1" cellspacing="3" cellpadding="3" border="1">
39
40                  <tbody>
41                      <tr>
42                          <th>
43                              Add new row,then test the delete function.
44                          </th>
45                      </tr>
46                  </tbody>
47              </table>
48          </div>
49      </body>
50  </html>
```

7.3.3 使用 JQuery 生成精美的 Tab 按钮

Tab 是制作网页过程中常用的控件，虽然在 JavaScript 中也有这样的按钮控件，但是不够精美，我们要完成的是较为精美的 Tab 按钮，如图 7.4 所示。

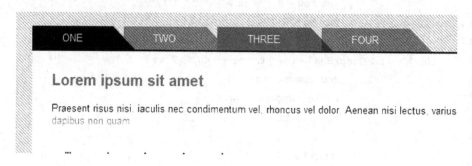

图 7.4 Tab 页面样式

要完成图 7.4 所示的效果，需要的内容如下。

1）编写相应的 CSS 样式。Tab 内部的文字颜色、背景、透明度条纹等都需要使用 CSS 样式来编写。

2）设计页面整体的布局，使得这个布局整洁、大气。

3）编写对应的 Function，包括背景的转换、Tab 的切换、文字的切换、分部页面的连接等。

页面代码如下所示。

```
1      <!DOCTYPE html>
2      <!-- saved from url=(0105) http://red-team-design.com/wp-content/
3      uploads/2012/05/google-play-minimal-tabs-with-css3-jquery-demo.html -->
4      <html><head><meta http-equiv="Content-Type" content="text/html;
5      charset=UTF-8">
6      <title></title>
7      <style>
8      body {
9        width: 700px;
10       margin: 100px auto 0 auto;
11       font-family: Arial, Helvetica;
12       font-size: small;
13       background-color: #eee;
14       background-image: url(data:image/gif;base64,R0lGODlhCAAIAJEAAMzMzP//////
15     wAAACH5BAEHAAIALAAAAAAIAAgAAAAINhG4nudroGJBRsYcxKAA7);
16     }
17     #tabs {
18       overflow: hidden;
19       width: 100%;
20       margin: 0;
```

```css
21        padding: 0;
22        list-style: none;
23      }
24      #tabs li {
25        float: left;
26        margin: 0 -15px 0 0;
27      }
28      #tabs a {
29        float: left;
30        position: relative;
31        padding: 0 40px;
32        height: 0;
33        line-height: 30px;
34        text-transform: uppercase;
35        text-decoration: none;
36        color: #fff;
37        border-right: 30px solid transparent;
38        border-bottom: 30px solid #3D3D3D;
39        border-bottom-color: #777 \9;
40        opacity: .3;
41        filter: alpha(opacity=30);
42      }
43      #tabs a:hover,
44      #tabs a:focus {
45        border-bottom-color: #2ac7e1;
46        opacity: 1;
47        filter: alpha(opacity=100);
48      }
49      #tabs a:focus {
50        outline: 0;
51      }
52      #tabs #current {
53        z-index: 3;
54        border-bottom-color: #3d3d3d;
55        opacity: 1;
56        filter: alpha(opacity=100);
57      }
58      /* ----------- */
59      #content {
60         background: #fff;
61         border-top: 2px solid #3d3d3d;
62         padding: 2em;
63         /* height: 220px;*/
```

```
64              }
65              #content h2,
66                #content h3,
67                #content p {
68                  margin: 0 0 15px 0;
69              }
70              /* Demo page only */
71              #about {
72                  color: #999;
73                  text-align: center;
74                  font: 0.9em Arial, Helvetica;
75              }
76              #about a {
77                  color: #777;
78              }
79          </style>
80          <script async="" src="./Google Play's minimal tabs with CSS3 & jQuery-
81   demo_files/adpacks-demo.js">
82          </script>
83          <style type="text/css">#adpacks-wrapper{font-family:Arial,Helvetica;width:
84   280px;position:fixed;_position:absolute;bottom:0;right:5px;z-index:9999;background:
85   #eee;padding:15px;box-shadow:0 0 2px #444;border-radius:5px 5px 0 0;}body.adpacks{back-
86   ground:#fff;padding:15px;margin:15px 0 0;border:3px solid #eee;}body.one.bsa_it_ad
87   {background:transparent;border:none;font-family:inherit;padding:0;margin:0;}body.one
88   .bsa_it_ad.bsa_it_i{display:block;padding:0;float:left;margin:0 10px 0 0;}body
89   .one.bsa_it_ad.bsa_it_i img{padding:0;border:none;}body.one.bsa_it_ad.bsa_
90   it_t{padding:0 0 6px 0;font-size:11px;}body.one.bsa_it_p{display:none;}body #bsap
91   _aplink,body #bsap_aplink:hover{display:block;font-size:9px;margin:-20px 0 0 0;text-a-
92   lign:right;}body.one.bsa_it_ad.bsa_it_d{font-size:11px;}body.one{overflow:hid-
93   den}
94          </style>
95          <script type="text/javascript" async="" src="./Google Play's minimal
96   tabs with CSS3 & jQuery-demo_files/bsa.js"></script><script type="text/
97   javascript" id="_bsap_js_a5f348233fceef06ba365ab392938d10" src="./
98   Google Play's minimal tabs with CSS3 & jQuery-demo_files/s_
99   a5f348233fceef06ba365ab392938d10.js" async="async">
100         </script>
101         <style type="text/css" id="bsa_css">.one{position:relative}.one.bsa_it_
102  ad{display:block;padding:15px;border:1px solid #e1e1e1;background:#f9f9f9;font-
103  family:helvetica,arial,sans-serif;line-height:100%;position:relative}.one.bsa_
104  it_ad a{text-decoration:none}.one.bsa_it_ad a:hover{text-decoration:none}.one
105  .bsa_it_ad.bsa_it_t{display:block;font-size:12px;font weight:bold;color:#
106  212121;line-height:125%;padding:0 0 5px 0}.one.bsa_it_ad.bsa_it_d{display:
107  block;font-size:11px;color:#434343;font-size:12px;line height:135%}.one.bsa_it_
108  ad.bsa_it_i{float:left;margin:0 15px 10px 0}.one.bsa_it_p{display:block;text-
109  align:right;position:absolute;bottom:10px;right:15px}.one.bsa_it_p a{font-size:
110  10px;color:#666;text-decoration:none}.one.bsa_it_ad.bsa_it_p a:hover{font-
111  style:italic}
```

```html
112        </style>
113      </head>
114      <body>
115      <ul id="tabs">
116        <li><a href="http://red-team-design.com/wp-content/uploads/2012/05/
117   google-play-minimal-tabs-with-css3-jquery-demo.html#" name="#tab1" id="cur-
118   rent">One</a></li>
119        <li><a href="http://red-team-design.com/wp-content/uploads/2012/05/google-play-
120   minimal-tabs-with-css3-jquery-demo.html#" name="#tab2" id="">Two</a></li>
121        <li><a href="http://red-team-design.com/wp-content/uploads/2012/05/
122   google-play-minimal-tabs-with-css3-jquery-demo.html#" name="#tab3" id="">
123   Three</a></li>
124        <li><a href="http://red-team-design.com/wp-content/uploads/2012/05/
125   google-play-minimal-tabs-with-css3-jquery-demo.html#" name="#tab4" id="">
126   Four</a></li>
127      </ul>
128      <div id="content">
129      <div id="tab1" style="display: block;">
130      <h2>Lorem ipsumsit amet</h2>
131      <p>Praesent risus nisi, iaculis nec condimentum vel, rhoncus vel do lor.
132   Aenean nisi lectus, varius nec tempus id, dapibus non quam.</p>
133      <p>Suspendisse ac libero mauris. Cras lacinia porttitor urna, vitae mol-
134   estie libero posuere et. Mauris turpis tortor, mollis non vulputate sit amet,
135   rhoncus vitae purus.</p>
136      <h3>Pellentesque habitant</h3>
137      <p>Vestibulum ante ipsum primis in faucibus orci luctus et ultrices posu-
138   ere cubilia curae.</p>
139      </div>
140      <div id="tab2" style="display: none;">
141      <h2>Vivamus fringilla suscipit justo</h2>
142      <p>Aenean dui nulla, egestas sit amet auctor vitae, facilisis id odio.
143   Donec dictum gravida feugiat.</p>
144      <p>Class aptent taciti sociosqu ad litora torquent per conubia nos tra,
145   per inceptos himenaeos. Cras pretium elit et erat condimentum et vo lutpat lo-
146   rem vehicula</p>
147      <p>Morbi tincidunt pharetra orci commodo molestie. Praesent ut leo nec
148   dolor tempor eleifend.</p>
149      </div>
150      <div id="tab3" style="display: none;">
151      <h2>Phasellus non nibh</h2>
152      <p>Non erat laoreet ullamcorper. Pellentesque magna metus, feugiat eu
153   elementum sit amet, cursus sed diam. Curabitur posuere porttitor lorem, eu ma-
154   lesuada tortor faucibus sed.</p>
```

```
155          <h3>Duis pulvinar nibh vel urna</h3>
156          <p>Donec purus leo, porttitor eu molestie quis, porttitor sit amet ipsum.
157     Class aptent taciti sociosqu ad litora torquent per conubia nostra, per incep-
158     tos himenaeos. Donec accumsan ornare elit id imperdiet.</p>
159          <p>Suspendisse ac libero mauris. Cras lacinia porttitor urna, vitae mol-
160     estie libero posuere et.</p>
161          </div>
162          <div id="tab4" style="display:none;">
163          <h2>Cum sociis natoque penatibus</h2>
164          <p>Magnis dis parturient montes, nascetur ridiculus mus. Nullam ac massa
165     quis nisi porta mollis venenatis sit amet urna. Ut in mauris velit, sed biben-
166     dum turpis.</p>
167          <p>Nam ornare vulputate risus, id volutpat elit porttitor non. In con se-
168     quat nisi vel lectus dapibus sodales. Pellentesque habitant morbi tris tique
169     senectus et netus et malesuada fames ac turpis egestas. Praesent bi bendum sag-
170     ittis libero.</p>
171          <h3>Imperdiet sem interdum nec</h3>
172          <p>Maurisrhoncus tincidunt libero quis fringilla.</p>
173          </div>
174          </div>
175          <p id="about">Back to <a href="http://red-team-design.com/google
176     play-minimal-tabs-with-css3-jquery">article</a>/ Drop me a message on <a
177     href="http://Weibo.com/catalinred">Weibo</a>!</p>
178          <script src="./Google Play's minimal tabs with CSS3 & jQuery-demo_
179     files/jquery-1.7.2.min.js"></script>
180          <script>
181              function resetTabs(){
182                  $("#content>div").hide(); //Hide all content
183                  $("#tabs a").attr("id",""); //Reset id's
184              }
185              var myUrl = window.location.href; //get URL
186              var myUrlTab = myUrl.substring(myUrl.indexOf("#")); // Forlocal host/
187     tabs.html#tab2, myUrlTab = #tab2
188              var myUrlTabName = myUrlTab.substring(0,4); // For the above exam ple,
189     myUrlTabName = #tab
190              (function(){
191                  $("#content>div").hide(); // Initially hide all content
192                  $("#tabs li:first a").attr("id","current"); // Activate first tab
193                  $("#content>div:first").fadeIn(); // Show first tab content
194                  $("#tabs a").on("click",function(e){
195                      e.preventDefault();
196                      if($(this).attr("id") == "current"){ //detection for cur
197     rent tab
```

```
198                    return
199                }
200                else{
201                    resetTabs();
202                    $(this).attr("id","current"); // Activate this
203                    $($(this).attr('name')).fadeIn(); // Show content for current tab
204                }
205            });
206            for (i=1; i<=$("#tabs li").length; i++) {
207                if (myUrlTab==myUrlTabName + i) {
208                    resetTabs();
209                    $("a[name='" + myUrlTab + "']").attr("id","current"); //
210    Activate url tab
211                    $(myUrlTab).fadeIn(); // Show url tab content
212                }
214            }
215        })()
216    </script>
217    <!--BSA AdPacks code-->
218    <script>
219        (function(){
220        var bsa=document.createElement('script');
221          bsa.async=true;
222          bsa.src='http://www.red-team-design.com/js/adpacks-demo.js';
223        (document.getElementsByTagName('head')[0]||document.getElements
224    ByTagName('body')[0]).appendChild(bsa);
225        })();
226    </script>
227    <div id="adpacks-wrapper">
228    <div id="bsap_1257097" class="bsap_1257097 bsap">
229    <div class="bsa_it one">
230    <div class="bsa_it_ad ad1 odd" id="bsa_4452407">
231    <a href=" http://stats.buysellads.com/click.go?z=1257097&b=
232    4452407&g=&s=&sw=1440&sh=900&br=chrome,33,win&r=
233    0.9319815100170672&link=http://bit.ly/MIg8Dn" onmouseover="win dow.status
234    =&#39;http://bit.ly/MIg8Dn&#39;; return true;" onmouseout="window.status
235    =&#39;&#39;; return true;" target="_blank">
236        <span class="bsa_it_i">
237        <img src="./Google Play's minimal tabs with CSS3 & jQuery-demo_files/GT-
238    NDC27ICWAI627MCW7LYK7UCKBILKJIF67DCZ3ICA7IP5QLCVAILK3KHAUN4AS
239    IKMBFCZIIKM7FK7Z6KQALBYIMGH7NPSDKKHLFCZID2RNLOA3G2MENAYZFKWUN4ASIKMB4
240    OBZ6HAAIP5QJ" width="130" height="100" alt="Custom Ecommerce">
241        </span>
```

```
242         </a>
243         <a href = " http://stats.buysellads.com/click.go? z = 1257097&b =
244         4452407&g = &s = &sw = 1440&sh = 900&br = chrome, 33, win&r =
245         0.25411732494831085&link = http://bit.ly/MIg8Dn" onmouseover = "win
246         dow.status=&#39;http://bit.ly/MIg8Dn&#39;;return true;" onmouseout = "win-
247         dow.status=&#39;&#39;;return true;" target = "_blank">
248         <span class = "bsa_it_t">Custom Ecommerce
249         </span>
250         <span class = "bsa_it_d">Easy, unlimited, affordable. Create a free ac-
251         count today.
252         </span>
253         </a>
254         <div style = "clear:both"></div></div><span class = "bsa_it_p">
255         <a href = "http://buysellads.com/buy/detail/40406/zone/1257097?utm_
256         source=site_40406&utm_medium=website&utm_campaign=imagetext&utm_con-
257         tent=zone_1257097">ads by BSA</a></span></div></div>
258         <a href="http://adpacks.com/" id = "bsap_aplink">via Ad Packs</a>
259         </div>
260         </body>
261         </html>
```

更多精美的 Tab 空间设计内容，可参见 http://www.cnblogs.com/lhb25/archive/2012/11/26/10-useful-jquery-tab-plugins.html。

7.3.4 使用 JQuery 完成相框效果

相框效果是在网页上常常用到的一种动态效果，它首先把我们上传的照片做成缩略图进行排列，当鼠标移动到页面其中的某一个位置时就显示相应的照片；当单击照片时，对应照片放大以显示大图，如图 7.5 所示。代码范例如下所示。

图 7.5 相框效果示例

存放图片的代码如下。

```
1        <ul>
2            <li>
3                <img src = "/source.jpg" alt = "" />
4                <p>Caption Here</p>
5            </li>
6            ...
7        </ul>
```

CSS 样式如下。

```
1    #pg { position: relative; height: 585px; background: #000; }
2    #pg li { position: relative; list-style: none; width: 175px; height: 117px;
3    overflow: hidden; float: left; z-index: 2; opacity: .3; }
4    #pg li.active { opacity: 1; }
5    #pg li.selected { opacity: 1; z-index: 99;-moz-box-shadow: 0px 0px 10px #
6    fff;-webkit-box-shadow: 0px 0px 10px #fff; }
7    #pg li img { display: block; width: 100% ; }
8    #pg li p { color: white; margin: 10px 0; font-size: 12px; }
```

Function 代码如下。

```
1    $('#pg').jphotogrid({
2        baseCSS: {
3            width: '175px',
4            height: '117px',
5            padding: '0px'
6        },
7        selectedCSS: {
8            top: '50px',
9            left: '100px',
10           width: '500px',
11           height: '360px',
12           padding: '10px'
13       }
14   });
```

※习 题

1. 使用 JQuery 实现放大镜效果，效果如图 7.6 所示。
2. 使用 JQuery 实现页面上图片的放大和隐藏功能。
3. 编写一个动态菜单，形式如图 7.7 所示。
4. 设计滚动条空间，要求：
1) 推动滚动条时，图片跟着滚动。
2) 单击图片时，图片突出显示，滚动条也移动到单击的图片位置，样式如图 7.8 所示。

图 7.6 放大镜效果[①]

图 7.7 动态菜单

图 7.8 滚动条空间

※综合应用

1. 制作一个使用 JQuery 完成的动态广告。广告里面共有三款手机，交替显示，一个手

---

① 可参见 http：//www.17sucai.com/pins/358.html 和 http：//www.starplugins.com/cloudzoom。

机旋转进入，一个手机水平切入，一个手机竖直切入，分别如图7.9～图7.11所示。

图7.9 竖直切入的手机

图7.10 旋转进入的手机

图7.11 水平切入的手机

页面主代码如下所示。

```
1    <div id="ca_banner1" class="ca_banner ca_banner1">
2        <div class="ca_slide ca_bg1">
3            <div class="ca__zone ca_zone1"><!--Product-->
4                <div class="ca_wrap ca_wrap1">
5                    <img src="images/product1.png"class="ca_shown"alt=""/>
6                    <img src="images/product2.png"alt=""style="display:none;"/>
7                    <img src="images/product3.png"alt=""style="display:none;"/>
```

```
8              <img src="images/product4.png"alt=""style="display:none;"/>
9              <img src="images/product5.png"alt=""style="display:none;"/>
10         </div>
11      </div>
12      </div class="ca_zone ca_zone2"><!--Line-->
13         <div class="ca_wrap ca_wrap2">
14             <img src="images/line1.png"class="ca_shown" alt=""/>
15             <img src="images/line2.png"alt=""style="display:none;"/>
16         </div>
17      </div>
18      <div class="ca_zone ca_zone3"><!--Title-->
19         <div class="ca_wrap ca_wrap3">
20             <img src="images/title1.png"class="ca_shown"alt=""/>
21             <img src="images/title2.png"alt=""style="display:none;"/>
22             <img src="images/title3.png"alt=""style="display:none;"/>
23         </div>
24      </div>
25    </div>
26 </div>
```

主要CSS样式如下所示。

```
1   .ca_banner{
2       position:relative;
3       overflow:hidden;
4       background:#f0f0f0;
5       padding:10px;
6       border:1px solid #fff;
7       -moz-box-shadow:0px 0px 2px #aaa inset;
8   }
9   .ca_slide{
10      width:100%
11      height:100%
12      position:relative;
13      overflow:hidden;
14  }
15  .ca_zone{
16      position:absolute;
17      width:100% ;
18  }
19  .ca_wrap{
20      position:relative;
21      display:table-cell;
22      vertical-align:middle;
23      text-align:center;
```

```css
24      }
25      ...
26      .ca_banner1.ca_wrap3{
27          width:378px;
28          height:31px;
29      }
```

主要JQuery代码如下所示。

```javascript
1     $('#ca_banne1').banner({
2         steps:[
3             [
4                 //1 step:
5                 [{"to":"2"},{"effect":"zoomOutRotated-zoomInRotated"}],
6                 [{"to":"1"},{}],
7                 [{"to":"2"},{"effect":"slideOutRight-slideInRight"}]
8             ],
9             [
10                //2 step:
11                [{"to":"3"},{"effect":"slideOutTop-slideInTop"}],
12                [{"to":"1"},{}],
13                [{"to":"2"},{}]
14            ],
15            [
16                //3 step:
17                [{"to":"4"},{"effect":"zoomOut-zoomIn"}],
18                [{"to":"2"},{"effect":"slideOutRight-slideInRight"}],
19                [{"to":"2"},{}]
20            ],
21            [
22                 4 step
23                [{"to":"5"},{"effect":"slideOutBottom-slideInTop"}],
24                [{"to":"2"},{}],
25                [{"to":"3"},{"effect":"zoomOut-zoomIn"}]
26            ],
27            [
28                //5 step
29                [{"to":"1"},{"effect":"slideOutLeft-slideInLeft"}],
30                [{"to":"1"},{"effect":"zoomOut-zoomIn"}],
31                [{"to":"1"},{"effect":"slideOutRight-slideInRight"}]
32            ]
33        ],
34        total_steps:5,
35        speed:3000
36    });
```

2. 制作简单的 iPhone 解锁屏幕的滑动按钮，如图 7.12 所示。

图 7.12　解锁屏幕的滑动按钮

第 8 章

JQuery 高级应用

通过前面章节的学习，我们了解了什么是 JavaScript 和 JQuery，并初步使用了这两种前端技术。本章内容是对第 7 章的扩展与提高。在本章中，我们将介绍用 JQuery 编写的控件：zTree、JQGrid，学习这些树形控件和表格控件的调用方式、语句格式、作用和使用它们可以达到什么样的效果等内容。这些控件使用 AJAX 技术，支持异步刷新，可以很好地支持前台程序的开发。

8.1 zTree 控件

zTree 是一个依靠 JQuery 实现的多功能"树插件"。优异的性能、灵活的配置、多种功能的组合是 zTree 的优点。它专门适合项目开发，尤其适用于树状菜单、树状数据的 Web 显示、权限管理等应用。

zTree 是开源免费的软件（有 MIT 许可证）。在开源的推动下，zTree 越来越完善，目前已经拥有了不少使用者，并且今后还会推出更多的 zTree 扩展功能库，让 zTree 更加强大。目前 zTree 的优点在于：

1）有专门的团队在维护并不断升级 zTree，使得 zTree 这一控件在不断完善并得到优化（目前最新版本是 v3.5.16）。

2）从 3.0 版本开始，zTree 的核心代码按照功能进行了分割，使得不需要使用的代码不用加载，大大提高了控件的运行速度。

3）采用了延迟加载技术，上万节点可轻松加载，即使在 IE 6 下也能基本做到轻松加载（由于浏览器升级越来越快，因此以后的版本可能不再支持 IE 6 版本）。

4）兼容 IE、FireFox、Chrome、Opera、Safari 等浏览器。

5）支持静态和 AJAX 异步加载节点数据。

6）通过 CSS 技术实现任意更换皮肤 / 自定义图标的功能。

7）支持极其灵活的 checkbox 或 radio 选择功能。

8）提供多种事件响应回调。

9）灵活的编辑（增加/删除/修改/查询）功能，可随意拖拽节点，还可以多节点拖拽。

10）在一个页面内可同时生成多个 Tree 实例。

11）简单的参数配置可实现灵活多变的功能。

zTree 是基于 CSS 和 JQuery 的控件，在使用前需要引入相应的 JQuery 包和 CSS 样式模板。目前，zTree 可以在 http：//www.ztree.me/v3/main.php#_zTreeInfo 上进行下载，解压缩后将里面的 js 文件夹的内容和 css 文件夹的内容复制到项目对应的文件夹下面，包的引用格式如下所示：

```
1    <link rel="stylesheet" href="../../../css/zTreeStyle/zTreeStyle.css"
2    type="text/css">
3    <script type="text/javascript" src="../../../js/jquery1.4.4.min.js">
4    </script>
5    <script type="text/javascript" src="../../../js/jquery.ztree.core
6    3.5.js"></script>
```

在这里，需要引用 zTreeStyle.css、jquery-1.4.4.min.js 和 jquery.ztree.core-3.5.js 三个包。目前 zTree 使用 jquery-1.4.4 作为默认的 JQuery 包，但是也可以使用更高版本的 JQuery 来支持（需要注意的是，目前并不是所有版本的 JQuery 都可以很好地支持 zTree，笔者目前测试的最高版本是 jquery-1.8.3，可以很好地支持 zTree）。如果想要改变 zTree 中图标、选项框等的样式，可以在 zTreeStyle.css 中进行修改。这三个包是使用 zTree 时需要添加的最为基本的引用，其他功能需要的引用，笔者会在下文中一一介绍。

引用包之后，我们需要对 zTree 进行插入节点操作，这样才能最终完成树的建立。其中，插入节点时需要注意树上节点的父子关系。在 zTree 中，父子节点关系定义如下所示。

```
1    var nodes = [
2        {name:"父节点1", children:[
3            {name:"子节点1"},
4            {name:"子节点2"}
5        ]}
6    ];
```

【例 8.1】 基本的 zTree 树的建立。下面建立如图 8.1 所示的树。

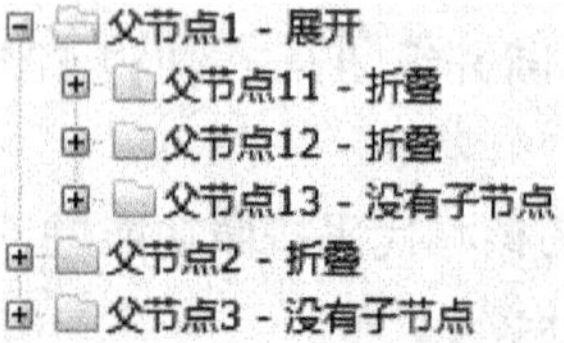

```
父节点1 - 展开
    父节点11 - 折叠
    父节点12 - 折叠
    父节点13 - 没有子节点
父节点2 - 折叠
父节点3 - 没有子节点
```

图 8.1　zTree 结果图

要求：在初始化时，要求父节点 1 是展开的，父节点 2 和父节点 3 是折叠的，且父节点 2 下面有下一级子树和叶子节点，而父节点 3 是没有子节点的。本例的代码如下所示：

```
1    <HEAD>
2    <link rel="stylesheet" href="../../../css/zTreeStyle/zTreeStyle.css"
3    type="text/css">
4    <script type="text/javascript" src="../../../js/jquery-1.4.4.min.js">
5    </script>
6    <script type="text/javascript" src="../../../js/jquery.ztree.core-
7    3.5.js"></script>
```

```
8     <SCRIPT type="text/javascript">
9         var setting = {};
10        var zNodes =[
11            { name:"父节点1 - 展开", open:true,
12                children: [
13                    { name:"父节点11 - 折叠",
14                        children: [
15                            { name:"叶子节点111"},
16                            { name:"叶子节点112"},
17                            { name:"叶子节点113"},
18                            { name:"叶子节点114"}
19                        ]},
20                    { name:"父节点12 - 折叠",
21                        children: [
22                            { name:"叶子节点121"},
23                            { name:"叶子节点122"},
24                            { name:"叶子节点123"},
25                            { name:"叶子节点124"}
26                        ]},
27                    { name:"父节点13 - 没有子节点", isParent:true}
28                ]},
29            { name:"父节点2 - 折叠",
30                children: [
31                    { name:"父节点21 - 展开", open:true,
32                        children: [
33                            { name:"叶子节点211"},
34                            { name:"叶子节点212"},
35                            { name:"叶子节点213"},
36                            { name:"叶子节点214"}
37                        ]},
38                    { name:"父节点22 - 折叠",
39                        children: [
40                            { name:"叶子节点221"},
41                            { name:"叶子节点222"},
42                            { name:"叶子节点223"},
43                            { name:"叶子节点224"}
44                        ]},
45                    { name:"父节点23 - 折叠",
46                        children: [
47                            { name:"叶子节点231"},
48                            { name:"叶子节点232"},
49                            { name:"叶子节点233"},
50                            { name:"叶子节点234"}
```

```
51                        ]}
52                    ]},
53                { name:"父节点 3 - 没有子节点", isParent:true}
54            ];
55            $(document).ready(function(){
56                $.fn.zTree.init($("#treeDemo"), setting, zNodes);
57            });
58        </SCRIPT>
59    </HEAD>
60    <BODY>
61        <div class = "zTreeDemoBackground left">
62            <ul id = "treeDemo" class = "ztree"></ul>
63        </div>
64    </BODY>
```

其中，zNodes 是所有的节点集合，父子关系的代码模式可以参见上文中父子节点关系定义代码。"open：true"是设置树节点初始显示时是否展开，其中 true 表示初始时为展开，false 表示初始时为折叠，默认为 false。本例中因为需要"父节点 1"初始时为展开，所以设置"open：true"。由于"父节点 3"没有子节点，但又需要将其设定为父节点而不是叶子节点，因此需要设置属性"isParent"。当"isParent"属性为 true 时，无论节点下面是否有子节点，本节点都将设置为父节点，否则按照节点下面是否有子节点来判断。这里"父节点 3"需要添加属性"isParent：true"。最后，使用"$.fn.zTree.init($("#treeDemo"), setting, zNodes);"将 setting 和 zNodes 添加到树里面。这里，treeDemo 是 <body></body> 里面的一对 <ul></ul> 的 id（ul 是 tree 的载体）。setting 将会在下文中介绍。

8.2　zTree 的 API

8.2.1　API 综述

zTree 的 API 主要分为三类：setting 配置、zTree 方法和 treeNode 节点数据。总体方法如下所示。

setting 配置：

```
1     var setting = {
2         treeId : "",
3         treeobj : null,
4         async : {
5             autoParam : [];
6             contentType : "application…"
7             dataFilter : null,
8             dataType : "text",
9             enable : false,
10            otherParam : [],
```

```
11                type : "post",
12                url : ""
13            },
14        Callback : {
15                befoerAsync : null,
16                beforeCheck : null,
17                beforeClick : null,
18                beforeCollapse : null,
19                beforeDblClick : null,
20                beforeDrag : null,
21                beforeDragOpen : null,
22                beforeDrop : null,
23                beforeEditName : null,
24                beforeExpand : null,
25                beforeMouseDown : null,
26                beforeMouseUp : null,
27                beforeRemove : null,
28                beforeRename : null,
29                beforeRightClick : null,
30                onAsyncError : null,
31                onAsyncSuccess : null,
32                onCheck : null,
33                onClick : null,
34                onCollapse : null,
35                onDblClick : null,
36                onDrag : null,
37                onDragMove : null,
38                onDrop : null,
39                onExpand : null,
40                onMouseDown : null,
41                onMouseUp : null,
42                onNodeCreated : null,
43                onRemove : null,
44                onRename : null,
45                onRightClick : null
46            },
47        Check : {
48                autoCheckTrigger : false,
49                chkboxType : {"Y": "ps", "N": "ps"},
50                chkStyle : "checkbox",
51                enable : false,
52                nocheckInherit : false,
53                chkDisabledInherit : false,
```

```
54              radioType : "level"
55          },
56      Data : {
57          keep : {
58              leaf : false,
59              parent : false
60          },
61          Key : {
62              Checked : "checked",
63              Children : "children",
64              Name : "name",
65              Title : "",
66              url : ""
67          },
68          simpleData : {
69              enable : false,
70              idKey : "id",
71              pIdKey : "pId",
72              rootPid : null
73          }
74      },
75      Edit : {
76          Drag : {
77              autoExpandTrigger : true,
78              isCopy : true,
79              isMove : true,
80              prev : true,
81              next : true,
82              inner : true,
83              borderMax : 10,
84              borderMin : -5,
85              minMoveSize : 5,
86              maxShowNodeNum : 5,
87              autoOpenTime : 500
88          },
89          editNameSelectAll : false,
90          enable : false,
91          removeTitle : "remove",
92          renameTitle : "rename",
93          showRemoveBtn : true,
94          showRenameBtn : true
95      },
96      View : {
```

```
97              addDiyDom : null,
98              addHoverDom : null,
99              autoCancelSelected : true,
100             dblClickExpand : true,
101             expandSpeed : "fast",
102             fontCss : {},
103             nameIsHTML : false,
104             removeHoverDom : null,
105             selectedMulti : true,
106             showIcon : true,
107             showLine : true,
108             showTitle : true,
109             txtSelectedEnable : false
110         }
111     }
```

zTree 方法：

```
1       $.fn.zTree : {
2           Init (obj, zSetting, zNodes)
3           getZTreeObj (treeId)
4           destroy (treeId)
5           _z : (tools, view, enent, data)
6       }
7       zTreeObj : {
8           setting
9           addNodes(parentNode, newNodes, isSilent)
10          cancelEditName(newName)
11          cancelSelectedNode(node)
12          checkAllNodes(checked)
13          checkNode(node, checked, checkTypeFlag, callbackFlag)
14          copyNode(targetNode, node, moveType, isSilent)
15          destroy()
16          editName(node)
17          expandAll(expandFlag)
18          expandNode(node, expandFlag, sonSign, focus, callbackFlag)
19          getChangeCheckedNodes()
20          getCheckedNodes(checked)
21          getNodeByParam(key, balue, parentNode)
22          getNodeByTId(tId)
23          getNodeIndex(node)
24          getNodes()
25          getNodesByFilter(filter, isSingle, parentNode, invokeParam)
26          getNodesByParam(key, value, parentNode)
27          getNodesByParamFuzzy(key, value, jparentNode)
```

28		getSelectedNodes()
29		hideNode(node)
30		hideNodes(nodes)
31		moveNode(targetNode, node, moveType, isSilent)
32		reAsyncChildNodes(parentNode, reloadType, isSilent)
33		refresh()
34		removeChildNodes(parentNode)
35		removeNode(node, callbackFlag)
36		selectNode(node, addFlag)
37		setChkDisabled(node, disabled, inheritParent, inheritChildren)
38		setEditable(editable)
39		showNode(node)
40		showNodes(nodes)
41		transformToArray(nodes)
42		transformTozTreeNodes(simpleNodes)
43	updataNode(node, checkTypeFlag)	
44	}	

treeNode 节点数据：

1	treeNode : {	
2		checked
3		children
4		chkDisabled
5		click
6		getCheckStatus()
7		getNextNode()
8		getParentNode()
9		getPreNode()
10		halfCheck
11		icon
12		iconClose
13		iconOpen
14		iconSkin
15		inHidden
16		isParent
17		name
18		nocheck
19		open
20		target
21		url
22		*DIY*
23		[check_Child_State]
24		[check_Focus]
25		[checkedOld]

26		[editNameFlag]
27		[isAjaxing]
28		[isFirstNode]
29		[isHover]
30		[isLastNode]
31		[level]
32		[parentTId]
33		[tId]
34		[zAsync]
35	}	

在以上方法中，setting 配置用来对 zTree 上的操作进行配置，具体来说，就是用来设定如加载、单击、双击、右击等操作所触发的事件；zTree 方法用于配置 zTree 的初始参数、初始节点状态等信息；而 treeNode 节点数据用于对树上的节点进行参数配置。每种函数的具体用法将在下文中介绍。

8.2.2 常用 API 详解

1. setting 配置

setting 参数的格式如下所示。

```
1     var setting = {
2         treeId : "",
3         treeObj : null,
4         async : {
5             …
6         }
7         callback : {
8             …
9         }
10        check : {
11            …
12        }
13        data : {
14            …
15        }
16        …
17    }
```

（1）treeId

treeId 是 zTree 的唯一标识，初始化后，相当于用户定义的 zTree 容器的 id 属性值。注意，这个 id 属于内部参数，一般不做初始化与修改。

（2）treeObj

treeObj 是 zTree 容器的 JQuery 对象，主要好处是便于操作。

（3）async

async 是 zTree 的异步加载函数。zTree 在展开未展开过的节点时对其进行调用,用于加载此节点下的节点数据。以下参数都只在 setting. async. enable = true 时生效。

1) autoParam。

autoParam 为异步加载时需要自动提交父节点属性的参数,默认值为 [ ]。autoParam 将需要作为参数提交的属性名称制作成数组,例如:["id", "name"]。

【例 8.2】设置 autoParam 参数的提交。

① 设置 id 属性为自动提交的参数。

```
1        var setting = {
2            async: {
3                enable: true,
4                url: "../../Model/getStudent.cs",
5                autoParam: ["id"]
6            }
7        };
```

例如,在异步加载父节点(node = {id:1, name:" Father"})的子节点时,将提交参数 id = 1。

② 设置 id 属性为 FId,使之成为自动提交的参数。

```
1        var setting = {
2            async: {
3                enable: true,
4                url: "../../Model/getStudent.cs ",
5                autoParam: ["id = FId"]
6            }
7        };
```

例如,对父节点 node = {FId:1, name:"Father"}进行异步加载时,将提交参数 FId = 1。

③ 另外,亦可以将参数直接加到 url 后面进行传递。

```
1        var setting = {
2            async: {
3                enable: true,
4                url: "../../Model/getStudent.cs? id = " + FId,
5                autoParam: []
6            }
7        };
```

进行异步加载时,将提交参数 id = FId(这里,FId 是指 FId 里面存放的内容)。

2) contentType。

contentType 参数用于设置 AJAX 提交参数的数据类型,默认值:"application/x-www-form-urlencoded"。contentType = "application/x-www-form-urlencoded"按照标准的 Form 格式提交参数,可以满足绝大部分请求。contentType = " application/json" 按照 JSON 格式提交参数,可以满足. NET 编程的需要。

【例 8.3】设置 contentType 参数的提交。

下面设置提交参数数据类型为 JSON。

```
1            var setting = {
2                async: {
3                    enable: true,
4                    contentType: "application/json",
5                    url: "../../Model/getStudent.cs",
6                    autoParam: ["id"]
7                }
8            };
```

3) dataFilter。

dataFilter 是用于对异步加载的数据返回值进行预处理的函数,默认值为 null。

【例 8.4】设置 dataFilter 参数的提交。

下面设置异步加载后的预处理函数 AnnouncedDataReturn( )。

```
1            Function AnnouncedDataReturn(){
2                Alert("The data has been returned.");
3            }
4            var setting = {
5                async: {
6                    enable: true,
7                    contentType: "application/json",
8                    url: "../../Model/getStudent.cs",
9                    autoParam: ["id"]
10                   dataFilter: AnnouncedDataReturn
11               }
12           };
```

4) dataType。

dataType 用来获取的数据的类型,默认值是 "text"。

5) enable。

enable 用来设置是否开启异步加载模块,true 为开启模块,false 为关闭模块,默认值为 false。只有当 enable 为 true 时,以上介绍的方法才可以使用。

6) type。

type 用于设置 HTTP 请求的类型,有 "post" 和 "get" 两种,默认为 "post"。

7) url。

url 用于获取数据的 URL 地址。

(4) callback

callback 用于设置 zTree 上的触发事件,例如:异步加载前触发的事件,单击、双击前触发的事件,单击、双击时触发的事件,删除时触发的事件等。

callback 的格式为:

```
1            var setting = {
2                callback: {
3                    beforeAsync: BeforeASYNC,
4                    beforeClick: BeforeCLICK,
```

```
5              onClick: OnCLICK,
6              onRemove: OnREMOVE
7              …
8          }
9      };
```

事件名称后面紧跟着的是函数名称，如 BeforeASYNC、BeforeCLICK 等。当触发相应的事件时，将触发对应的函数。

2. zTree 方法

```
1      $.fn.zTree : {
2          init (obj, zSetting, zNodes)
3          getZTreeObj (treeId)
4          destroy (treeId)
5          _z : (tools, view, enent, data)
6      }
7      zTreeObj : {
8          setting
9          addNodes(parentNode, newNodes, isSilent)
10         expandAll(expandFlag)
11         ExpandNode(node, expandFlag, sonSign, focus, callbackFlag)
12         getChangeCheckedNotds(checked)
13         getCheckedNotds(checked)
14         getNodeByTId(tId)
15         getNodeIndex(node)
16         getNodes()
17         getSelectedNodes()
18         hideNode(node)
19         MoveNode(targetNode, treeNodenode, moveType, isSilent)
20         refresh()
21         removeNode(treeNodenode, callbackFlag)...
22     }
```

（1）init

init 是 zTree 的初始化方法，是必须使用的方法。其使用格式如下：

```
1      $(document).ready(function(){
2          zTreeObj = $.fn.zTree.init($("#tree"), setting, zTreeNodes);
3      });
```

在该方法中，将设定好的 setting 和父子关系存入 id 为 tree 的 ul 中。

（2）getZTreeObj

getZTreeObj 是用于根据 treeId 获得 zTree 对象的方法。注意，此方法必须在初始化 zTree 以后使用。默认时，treeId 即为存储 zTree 的 ul 的 id。例如：

var treeObj = $.fn.zTree.getZTreeObj("tree");

上述代码用于获取 id 为 tree 的 zTree 对象。

(3) destroy

destroy 是用于销毁 zTree 的方法，在 v3.4 以后的版本中才能使用。此方法可以销毁指定 treeId 的 zTree，也可以销毁当前页面全部的 zTree。销毁之后，重新使用已销毁的树时，必须使用 init 方法重新初始化。

【例 8.5】两种销毁树的方法。

1）销毁 id 为 "MustBeDestory" 的 zTree。

```
$.fn.zTree.destroy("MustBeDestory ");
```

2）销毁页面上所有的 zTree。

```
$.fn.zTree.destroy();
```

(4) addNodes

addNodes 方法用于添加节点，格式为 addNodes（parentNode, newNodes, isSilent）。其中，parentNode 用于指定父节点（使用这个节点的 id），如果要增加根节点，则设置 parentNode 为 null；NewNodes 是需要添加的节点数据 JSON 对象集合；isSilent 用来设置增加节点后其父节点是否直接展开。

【例 8.6】增加 3 个根节点。

对 id 为 "increaseNode" 的 zTree 增加 3 个节点。

```
1        var treeObj = $.fn.zTree.getZTreeObj("increaseNode");
2        var nNodes = [{name:"nNode1"}, {name:"nNode2"}, {name:"nNode3"}];
3        newNodes = treeObj.addNodes(null, nNodes);
```

(5) expandAll

expandAll（expandFlag）用于展开/折叠全部节点。若 expandFlag 为 true，则表示展开全部节点；若为 false，则表示折叠全部节点。

【例 8.7】展开树 tree1 的全部节点。

```
1        var treeObj = $.fn.zTree.getZTreeObj("tree1");
2        treeObj.expandAll(true)
```

(6) expandNode

expandNode（node, expandFlag, sonSign, focus, callbackFlag）用于展开/折叠指定的节点。其中，node 是需要展开/折叠的节点数据；expandFlag 表示展开还是折叠（true 为展开，false 为折叠）；sonSign 表示是展开所有子节点还是只展开自身（true 是全部展开，false 为展开自身）；focus 用于操作后设置焦点（true 表示展开/折叠后通过设置焦点保证此焦点进入可视区域内，false 表示不设置任何焦点），默认为 true；callbackFlag 为 true 表示执行此方法时触发 beforeExpand / onExpand 或 beforeCollapse / onCollapse 事件回调函数，否则不触发函数。

【例 8.8】展开树 tree1 中的当前选中节点的下属所有节点。

```
1        var treeObj = $.fn.zTree.getZTreeObj("tree1");
2        var nodes = treeObj.getSelectedNodes();
3        if (nodes.length > 0) {
4             treeObj.expandNode(nodes[0], true, true, true);
5        }
```

（7）getChangeCheckedNodes

getChangeCheckedNodes（）用于获取 checkbox 状态改变的节点集合，使用方法为：

```
1    var treeObj = $.fn.zTree.getZTreeObj("tree1");
2    var nodes = treeObj.getChangeCheckedNodes();
```

（8）getCheckedNodes

getCheckedNodes（checked）用于获取 checkbox 状态为 checked 的节点集合（若 checked 为 true，则表示获取被勾选的节点集合，若为 false，则表示获取未被勾选的节点集合）。使用方法为：

```
1    var treeObj = $.fn.zTree.getZTreeObj("tree1");
2    var nodes = treeObj.getCheckedNodes(true);
```

（9）getNodeByTId

getNodeByTId（tId）表示根据节点的 id 获取节点。

【例 8.9】 获取树 tree1 中节点 id 为 node1 的节点。

```
1    var treeObj = $.fn.zTree.getZTreeObj("tree1");
2    var node = treeObj.getNodeByTId("node1");
```

（10）getNodeIndex

getNodeIndex(node) 用于获取某节点在同级节点中的序号（从 0 开始）。

（11）getNodes

getNodes（）用于获取全部的节点。

（12）getSelectedNodes

getSelectedNodes（）用于获取 zTree 当前被选中的节点数据集合。获取的是当前节点及其子节点。

（13）hideNode

hideNode(node) 用于隐藏节点 node。

（14）hideNodes

hideNodes（）用于隐藏所有的节点。

（15）moveNodes

moveNodes（targetNode，treeNode，moveType，isSilent）用于移动节点。其中，targetNode 是指要移动到的目标节点，如果移动后成为根节点，设置 targetNode 为 null 即可；treeNode 是需要被移动的节点数据；moveType 指定移动到目标节点的相对位置，"inner"：成为子节点，"prev"：成为同级前一个节点，"next"：成为同级后一个节点；isSilent 设定移动节点后是否自动展开父节点，当 isSilent = true 时，不展开父节点，其他值或默认状态都自动展开。

【例 8.10】 移动树 tree1 中的节点。

1）将根节点中第 3 个节点移动到第 2 个节点前面。

```
1    var treeObj = $.fn.zTree.getZTreeObj("tree1");
2    var nodes = treeObj.getNodes();
3    treeObj.moveNode(nodes[1], nodes[2], "prev");
```

2）将根节点中第 3 个节点移动成为第 2 个节点的子节点。

```
1    var treeObj = $.fn.zTree.getZTreeObj("tree1");
2    var nodes = treeObj.getNodes();
3    treeObj.moveNode(nodes[1], nodes[2], "inner");
```

（16）refresh

refresh()用于刷新树。

（17）removeNode

removeNode（treeNode，callbackFlag）用于删除节点。其中，treeNode 是需要删除的节点数据；若 callbackFlag 为 true，则表示执行此方法时触发 beforeRemove 或 onRemove 事件回调函数，若为 false，则表示执行此方法时不触发事件回调函数。

【例 8.11】删除树 tree1 中选中的第 2 个节点。

```
1    var treeObj = $.fn.zTree.getZTreeObj("tree");
2    var nodes = treeObj.getSelectedNodes();
3    treeObj.removeNode(nodes[1]);
```

3. treeNode 节点数据

treeNode 节点数据部分方法如下。

```
1    treeNode : {
2        children
3        click
4        getParentNode()
5        icon
6        isParent
7        url
8        ...
9    }
```

（1）children

children 是节点的子节点的数据集合。如果不使用 children 属性保存子节点数据，则需要修改 setting.data.key.children；异步加载时，对于设置了 isParent = true 的节点，在展开时将进行异步加载。

（2）click

最简单的 click 事件操作。相当于 onclick = "..." 的内容。语法样例如下：

```
1    var nodes = [
2        {"id":1, "name":"nam1", "url":"../login/login.xhtml", "click":"a-
3    lert('test');"},
4        ...
5    ]
```

（3）getParentNode

getParentNode()用于获取 treeNode 节点的父节点，语法样例如下。

```
1    var treeObj = $.fn.zTree.getZTreeObj("tree1");
2    var sNodes = treeObj.getSelectedNodes();
3    var node = sNodes[0].getParentNode();
```

（4）icon

icon 用于自定义树前面的图片，语法样例如下。

```
1    var nodes = [
2        { name:"parent1", icon:"../img/parent.gif"}}
3    ]
```

（5）isParent

isParent 记录 treeNode 节点是否为父节点。如果返回值是 true，则表示本节点是父节点，若为 false，则表示不是父节点。语法样例如下。

```
1    var treeObj = $.fn.zTree.getZTreeObj("tree");
2    var sNodes = treeObj.getSelectedNodes();
3    var isParent = sNodes[0].isParent;
```

（6）url

url 存储节点连接的目标，语法样例如下。

```
1    var nodes = [
2        { "id":1, "name":"nam1", "url":"../login/login.xhtml", "click":"alert('
3        test');"},
4        …
5    ]
```

8.3 zTree 应用实例

8.3.1 zTree 基本功能

1. 基本的 JSON 数据

建立一个 zTree，使用 JSON 来传递数据。

【例 8.12】使用 JSON 来传递数据，建立一个树，该树有 3 个父节点，父节点下有多个中间节点和叶子节点。该树的样式如图 8.2 所示。

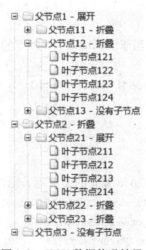

图 8.2 JSON 数据传递结果

建立这棵树的代码如下。

```
1    <link rel="stylesheet" href="../../../css/demo.css" type="text/css">
2    <link rel="stylesheet" href="../../../css/zTreeStyle/zTreeStyle.css"
3    type="text/css">
4    <script type="text/javascript" src="../../../js/jquery-1.4.4.min.js">
5    </script>
6    <script type="text/javascript" src="../../../js/jquery.ztree.core-
7    3.5.js"></script>
8    <SCRIPT type="text/javascript">
9        var setting = {
10           data: {
11               simpleData: {
12                   enable: true
13               }
14           }
15       };
16       var zNodes = [
17           { id:1, pId:0, name:"父节点1 - 展开", open:true},
18           { id:11, pId:1, name:"父节点11 - 折叠"},
19           { id:111, pId:11, name:"叶子节点111"},
20           { id:112, pId:11, name:"叶子节点112"},
21           { id:113, pId:11, name:"叶子节点113"},
22           { id:114, pId:11, name:"叶子节点114"},
23           { id:12, pId:1, name:"父节点12 - 折叠"},
24           { id:121, pId:12, name:"叶子节点121"},
25           { id:122, pId:12, name:"叶子节点122"},
26           { id:123, pId:12, name:"叶子节点123"},
27           { id:124, pId:12, name:"叶子节点124"},
28           { id:13, pId:1, name:"父节点13 - 没有子节点", isParent:true},
29           { id:2, pId:0, name:"父节点2 - 折叠"},
30           { id:21, pId:2, name:"父节点21 - 展开", open:true},
31           { id:211, pId:21, name:"叶子节点211"},
32           { id:212, pId:21, name:"叶子节点212"},
33           { id:213, pId:21, name:"叶子节点213"},
34           { id:214, pId:21, name:"叶子节点214"},
35           { id:22, pId:2, name:"父节点22 - 折叠"},
36           { id:221, pId:22, name:"叶子节点221"},
37           { id:222, pId:22, name:"叶子节点222"},
38           { id:223, pId:22, name:"叶子节点223"},
39           { id:224, pId:22, name:"叶子节点224"},
40           { id:23, pId:2, name:"父节点23 - 折叠"},
41           { id:231, pId:23, name:"叶子节点231"},
42           { id:232, pId:23, name:"叶子节点232"},
```

```
43                    { id:233, pId:23, name:"叶子节点233"},
44                    { id:234, pId:23, name:"叶子节点234"},
45                    { id:3, pId:0, name:"父节点3 - 没有子节点", isParent:true}
46              ];
47              $(document).ready(function(){
48                    $.fn.zTree.init($("#treeDemo"), setting, zNodes);
49              });
50        </SCRIPT>
51        <div class = "zTreeDemoBackground left">
52              <ul id = "treeDemo" class = "ztree"></ul>
53        </div>
```

2. 不显示连接线的树

下面通过一个示例介绍一下如何才能不显示树前面的连接线。

【例8.13】基于例8.12，建立可以不显示连接线的树，如图8.3所示。

图8.3 没有连接线的树

需要在setting中设置showLine值为false，即只需改变setting中的相应代码，其他部分的代码同例8.12。

```
1         var setting = {
2               view: {
3                     showLine: false
4               },
5               data: {
6                     simpleData: {
7                           enable: true
8                     }
9               }
10        };
```

3. 不显示节点图标的树

下面通过一个示例介绍一下如何才能不显示树节点前面的图标。

【例 8.14】 基于例 8.12，建立可以不显示非叶子节点的图标的树，如图 8.4 所示。

图 8.4 不显示非叶子节点的图标的树

需要在 setting 中设置 showIcon 为 showIcon，另外，通过 showIcon( ) 函数设定不显示非叶子节点的图标，即只需改变 setting 中的相应代码，其他部分的代码同例 8.12。

```
1    var setting = {
2        view: {
3            showIcon: showIcon
4        },
5        data: {
6            simpleData: {
7                enable: true
8            }
9        }
10    };
11    function showIcon(treeNode) {
12        return ! treeNode.isParent;
13    };
```

4. 自定义图标——icon 属性

下面通过一个示例介绍一下如何才能实现不同节点显示不同的图标。

【例 8.15】 建立不同节点显示不同图标的树，如图 8.5 所示。

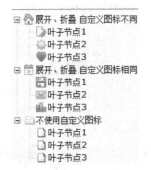

图 8.5 不同节点有不同的图标

需要对每个节点的icon属性进行设置，代码如下。

```
1   var setting = {
2       data: {
3           simpleData: {
4               enable: true
5           }
6       }
7   };
8   var zNodes = [
9       { id:1, pId:0, name:"展开、折叠自定义图标不同", open:true, iconOpen:"../../../css/zTreeStyle/img/diy/1_open.png", iconClose:"../../../css/zTreeStyle/img/diy/1_close.png"},
12      { id:11, pId:1, name:"叶子节点1", icon:"../../../css/zTreeStyle/img/diy/2.png"},
14      { id:12, pId:1, name:"叶子节点2", icon:"../../../css/zTreeStyle/img/diy/3.png"},
16      { id:13, pId:1, name:"叶子节点3", icon:"../../../css/zTreeStyle/img/diy/5.png"},
18      { id:2, pId:0, name:"展开、折叠自定义图标相同", open:true, icon:"../../../css/zTreeStyle/img/diy/4.png"},
20      { id:21, pId:2, name:"叶子节点1", icon:"../../../css/zTreeStyle/img/diy/6.png"},
22      { id:22, pId:2, name:"叶子节点2", icon:"../../../css/zTreeStyle/img/diy/7.png"},
24      { id:23, pId:2, name:"叶子节点3", icon:"../../../css/zTreeStyle/img/diy/8.png"},
26      { id:3, pId:0, name:"不使用自定义图标", open:true },
27      { id:31, pId:3, name:"叶子节点1"},
28      { id:32, pId:3, name:"叶子节点2"},
29      { id:33, pId:3, name:"叶子节点3"}
30  ];
```

5. 自定义文字字体

下面通过一个示例介绍一下如何实现不同节点显示不同的字体。

【例8.16】建立不同节点显示不同字体的树，如图8.6所示。

图8.6 不同节点显示不同字体的树

需要在每个节点设置其自身的font，并且为了让子节点可以继承父节点的字体，需要在setting中设置fontCss属性，代码如下所示。

```
1    var setting = {
2        view: {
3            fontCss: getFont,
4            nameIsHTML: true
5        }
6    };
7    var zNodes = [
8        { name:"粗体字", font:{'font-weight':'bold'} },
9        { name:"斜体字", font:{'font-style':'italic'}},
10       { name:"有背景的字", font:{'background-color':'black', 'color':'white
11   '}},
12       { name:"红色字", font:{'color':'red'}},
13       { name:"蓝色字", font:{'color':'blue'}},
14       { name:" <span style = 'color:blue;margin-right:0px;' >view </span>. <
15   span style = 'color:red;margin-right:0px;' >nameIsHTML </span>"},
16       { name:"zTree 默认样式"}
17   ];
18   function getFont(treeId, node) {
19       return node.font ? node.font : {};
20   }
```

6. 异步加载

下面的示例展示了各个子节点是如何采用异步加载方式的。

【例8.17】 建立异步加载的树。

异步加载需要在setting中使用async来设定异步加载的地址，使用asyncSuccess对返回后的数据进行处理，代码如下所示。

aspx 页面代码：

```
1    <%@ Page Language = "C#" AutoEventWireup = "true" CodeBehind = "Web
2    Form2.aspx.cs" Inherits = "CFBuilder.WebForm2" %>
3    <!DOCTYPE html PUBLIC "-//W3C//DTD XHTML 1.0 Transitional//EN" "http://
4    www.w3.org/TR/xhtml1/DTD/xhtml1-transitional.dtd">
5    <html xmlns = "http://www.w3.org/1999/xhtml">
6    <head runat = "server">
7    <title></title>
8    <link href = "Styles/zTreeStyle/zTreeStyle.css" rel = "stylesheet" type
9    = "text/css" />
10   <link href = "Styles/demo.css" rel = "stylesheet" type = "text/css" />
11   <script src = "Scripts/jquery-1.4.4.min.js" type = "text/javascript" ></
12   script>
13   <script src = "Scripts/jquery.ztree.core-3.5.min.js" type = "text/javas-
14   cript"></script>
```

```javascript
15      <script type="text/javascript">
16          var zNodes;
17          var zTree;
18          //setting 异步加载的设置
19          var setting = {
20              async: {
21                  enable:true, //表示异步加载生效
22                  url: "/AjaxService/WebData.aspx", //异步加载时访问的页面
23                  autoParam: ["id"], //异步加载时自动提交的父节点属性的参数
24                  otherParam:["ajaxMethod",'AnsyData'], //AJAX 请求时提交的
25  参数
26                  type:'post',
27                  dataType:'json'
28              },
29              checkable: true,
30              showIcon: true,
31              showLine: true, // zTree 显示连接线
32              data: {
33                  simpleData: {
34                      enable: true
35                  }
36              },
37              expandSpeed: "", //设置 zTree 节点展开、折叠的动画速度,默认
38  为"fast",""表示无动画
39              callback: { //回调函数
40                  onClick: zTreeOnClick, // 单击鼠标事件
41                  asyncSuccess: zTreeOnAsyncSuccess //异步加载成功事件
42              }
43          };
44          $(document).ready(function () {
45              Inint();
46              $.fn.zTree.init($("#treeDemo"), setting, zNodes);
47          });
48          //初始化加载节点
49          function Inint(){
50              $.ajax({
51                  url: '/AjaxService/WebData.aspx',
52                  type: 'post',
53                  dataType: 'json',
54                  async: false,
55                  data: { 'ajaxMethod': 'FirstAnsyData' },
56                  success: function (data) {
57                      zNodes = data;
```

```
58                              }
59                         });
60                    };
61                    //单击时获取 zTree 节点的 id 和 value 的值
62                    function zTreeOnClick(event, treeId, treeNode, clickFlag) {
63                         var treeValue = treeNode.id + "," + treeNode.name;
64                    };
65                    function zTreeOnAsyncSuccess(event, treeId, treeNode, msg) {
66                    }
67         </script>
68    </head>
69    <body>
70         <form id="form1" runat="server">
71             <div class="content_wrap">
72                 <div class="zTreeDemoBackground left">
73                     <ul id="treeDemo" class="ztree"></ul>
74                 </div>
75             </div>
76         </form>
77    </body>
78 </html>
```

异步提交的后台页面代码：

```
1  using System;
2  using System.Collections.Generic;
3  using System.Linq;
4  using System.Web;
5  using System.Web.UI;
6  using System.Web.UI.WebControls;
7  using Newtonsoft.Json;
8  using System.Data.SqlClient;
9  using System.Data;
10 namespace CFBuilder
11 {
12     public partial class WebData : System.Web.UI.Page
13     {
14         string strConn = @"Data Source=ANDY-PC\SQLEXPRESS;Initial
15 Catalog=Test;Integrated Security=True";
16         protected void Page_Load(object sender, EventArgs e)
17         {
18             #region
19             try
20             {
```

```csharp
21                        string ajaxMethod = Request["ajaxMethod"].ToString();//取
22        得前台AJAX请求的方法名称
23                        System.Reflection.MethodInfo method = this.GetType()
24        .GetMethod(ajaxMethod);
25                        if (method! =null)
26                        {
27                            method.Invoke(this, new object[] { });
28        //通过方法名称指向对应的方法
29                        }
30                    }
31                    catch (Exception)
32                    {
33                        throw;
34                    }
35                    finally
36                    {
37                        Response.End();
38                    }
39                    #endregion
40                }
41                public void AnsyData()
42                {
43                    List<object> lsNode = new List<object>();
44                    try
45                    {
46                        int id = int.Parse(Request.Params["id"]);
47                        using (SqlConnection conn = new SqlConnection(strConn))
48                        {
49                            string sql = "select * from OrginTree where OrgParent
50        =" + id + "";
51                            DataTable table = new DataTable();
52                            SqlDataAdapter dt = new SqlDataAdapter(sql, conn);
53                            dt.Fill(table);
54                            lsNode = getList(table);
55                            Response.Write(JsonConvert.SerializeObject(lsNode));
56                        }
57                    }
58                    catch (Exception)
59                    {
60                        throw;
61                    }
62                }
63                public bool isParentTrue(int ParentId)
```

```csharp
64              {
65                  try
66                  {
67                      using (SqlConnection conn = new SqlConnection(strConn))
68                      {
69                          conn.Open();
70                          string sql = "select * from OrginTree where OrgParent
71  = " + ParentId + "";
72                          DataTable table = new DataTable();
73                          SqlDataAdapter dt = new SqlDataAdapter(sql, conn);
74                          dt.Fill(table);
75                          return table.Rows.Count >= 1 ? true : false;
76                      }
77                  }
78                  catch (Exception)
79                  {
80                      throw;
81                  }
82              }
83              {
84                  try
85                  {
86                      TableEnjson tbEnjson = new TableEnjson();
87                      List<object> lsNode = new List<object>();
88                      using (SqlConnection conn = new SqlConnection(strConn))
89                      {
90                          conn.Open();
91                          string sql = "select * from OrginTree where OrgParent
92  is null";
93                          DataTable table = new DataTable();
94                          SqlDataAdapter dt = new SqlDataAdapter(sql, conn);
95                          dt.Fill(table);
96                          lsNode = getList(table);
97                          Response.Write(JsonConvert.SerializeObject(lsNode));
98  //用到了 Newtonsoft.dll 转化成 JSON 格式
99                      }
100                 }
101                 catch (Exception)
102                 {
103                     throw;
104                 }
105             }
106             public List<object> getList(DataTable table)
```

```csharp
107                 {
108                     try
109                     {
110                         List < object > lsNode = new List < object > ();
111                         bool isParent = true;
112                         foreach (DataRow row in table.Rows)
113                         {
114                             var ParentId = string.IsNullOrEmpty(row["OrgParent"].
115     ToString()) ? 0 : row["OrgParent"];
116                             if (isParentTrue(int.Parse(row["OrgId"].ToString())))
117                                 isParent = true;
118                             else
119                                 isParent = false;
120                             var zTreeData = new
121                             {
122                                 id = row["OrgId"],
123                                 pId = ParentId,
124                                 name = row["OrgName"],
125                                 isParent = isParent
126                             };
127                             lsNode.Add(zTreeData);
128                         }
129                         return lsNode;
130                     }
131                     catch (Exception)
132                     {
133                         throw;
134                     }
135                 }
136             }
```

数据库代码为：

```sql
1       CREATE TABLE OrginTree
2          (
3            OrgId INT PRIMARY KEY IDENTITY(1,1),
4            OrgName NVARCHAR(30),      //节点的名称
5            ORgParent INT              //父节点的 id
6          )
```

8.3.2 zTree 单选按钮/复选框功能

1. CheckBox 的勾选

下面的示例建立了一个 zTree，并展示了如何使用 CheckBox 对节点进行勾选。

【例 8.18】建立一棵含有 CheckBox 复选框的树，结果如图 8.7 所示。

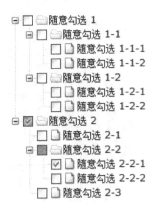

图 8.7 含有 CheckBox 复选框的树

若要设置此类树，则需要在 setting 中设置相应的 check 属性，代码如下所示。

```
1       var setting = {
2           check: {
3               enable: true
4           },
5           data: {
6               simpleData: {
7                   enable: true
8               }
9           }
10      };
11      var zNodes = [
12          { id:1, pId:0, name:"随意勾选 1", open:true},
13          { id:11, pId:1, name:"随意勾选 1-1", open:true},
14          { id:111, pId:11, name:"随意勾选 1-1-1"},
15          { id:112, pId:11, name:"随意勾选 1-1-2"},
16          { id:12, pId:1, name:"随意勾选 1-2", open:true},
17          { id:121, pId:12, name:"随意勾选 1-2-1"},
18          { id:122, pId:12, name:"随意勾选 1-2-2"},
19          { id:2, pId:0, name:"随意勾选 2", checked:true, open:true},
20          { id:21, pId:2, name:"随意勾选 2-1"},
21          { id:22, pId:2, name:"随意勾选 2-2", open:true},
22          { id:221, pId:22, name:"随意勾选 2-2-1", checked:true},
23          { id:222, pId:22, name:"随意勾选 2-2-2"},
24          { id:23, pId:2, name:"随意勾选 2-3"}
25      ];
```

2. Radio 的勾选

下面的示例建立了一个 zTree，并展示了如何使用 Radio 对节点进行勾选。

【例 8.19】建立一棵含有 Radio 单选按钮的树，结果如图 8.8 所示。

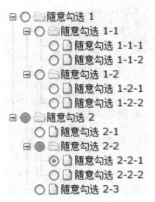

图 8.8　含有 Radio 单选按钮的树

若要设置此类树,则需要在 setting 中设置相应的 check 属性,代码如下所示。

```
1      var setting = {
2          check: {
3              enable: true,
4              chkStyle: "radio",
5              radioType: "level"
6          },
7          data: {
8              simpleData: {
9                  enable: true
10             }
11         }
12     };
13     var zNodes = [
14         { id:1, pId:0, name:"随意勾选 1", open:true},
15         { id:11, pId:1, name:"随意勾选 1-1", open:true},
16         { id:111, pId:11, name:"随意勾选 1-1-1"},
17         { id:112, pId:11, name:"随意勾选 1-1-2"},
18         { id:12, pId:1, name:"随意勾选 1-2", open:true},
19         { id:121, pId:12, name:"随意勾选 1-2-1"},
20         { id:122, pId:12, name:"随意勾选 1-2-2"},
21         { id:2, pId:0, name:"随意勾选 2", open:true},
22         { id:21, pId:2, name:"随意勾选 2-1"},
23         { id:22, pId:2, name:"随意勾选 2-2", open:true},
24         { id:221, pId:22, name:"随意勾选 2-2-1", checked:true},
25         { id:222, pId:22, name:"随意勾选 2-2-2"},
26         { id:23, pId:2, name:"随意勾选 2-3"}
27     ];
```

8.3.3 zTree 的拖拽功能

下面的示例建立了一个 zTree，并展示了如何对树上的节点进行拖拽操作。

【例 8.20】建立树，并把 1-2-1 节点拖拽到"随意拖拽 1"下，结果如图 8.9 所示。

图 8.9 拖拽结果展示

实现代码如下所示。

```
1   < link rel = "stylesheet" href = "../../../css/zTreeStyle/zTreeStyle.css" type
2   = "text/css" >
3       < script type = "text/javascript" src = "../../../js/jquery-1.4.4.min.js" > </
4   script >
5       < script type = "text/javascript" src = "../../../js/jquery.ztree.core
6   3.5.js" > </script >
7       < script type = "text/javascript" src = "../../../js/jquery.ztree.excheck
8   3.5.js" > </script >
9       < script type = "text/javascript" src = "../../../js/jquery.ztree.exedit
10  3.5.js" > </script >
11      < SCRIPT type = "text/javascript" >
12          var setting = {
13              edit: {
14                  enable: true,
15                  showRemoveBtn: false,
16                  showRenameBtn: false
17              },
18              data: {
19                  simpleData: {
20                      enable: true
21                  }
22              },
```

```
23              callback: {
24                  beforeDrag: beforeDrag,
25                  beforeDrop: beforeDrop
26              }
27          };
28          var zNodes = [
29              { id:1, pId:0, name:"随意拖拽 1", open:true},
30              { id:11, pId:1, name:"随意拖拽 1-1"},
31              { id:12, pId:1, name:"随意拖拽 1-2", open:true},
32              { id:121, pId:12, name:"随意拖拽 1-2-1"},
33              { id:122, pId:12, name:"随意拖拽 1-2-2"},
34              { id:123, pId:12, name:"随意拖拽 1-2-3"},
35              { id:13, pId:1, name:"禁止拖拽 1-3", open:true, drag:false},
36              { id:131, pId:13, name:"禁止拖拽 1-3-1", drag:false},
37              { id:132, pId:13, name:"禁止拖拽 1-3-2", drag:false},
38              { id:133, pId:13, name:"随意拖拽 1-3-3"},
39              { id:2, pId:0, name:"随意拖拽 2", open:true},
40              { id:21, pId:2, name:"随意拖拽 2-1"},
41              { id:22, pId:2, name:"禁止拖拽到我身上 2-2", open:true, drop:false},
42              { id:221, pId:22, name:"随意拖拽 2-2-1"},
43              { id:222, pId:22, name:"随意拖拽 2-2-2"},
44              { id:223, pId:22, name:"随意拖拽 2-2-3"},
45              { id:23, pId:2, name:"随意拖拽 2-3"}
46          ];
47          function beforeDrag(treeId, treeNodes) {
48              for (var i=0,l=treeNodes.length; i<l; i++) {
49                  if (treeNodes[i].drag === false) {
50                      return false;
51                  }
52              }
53              return true;
54          }
55          function beforeDrop(treeId, treeNodes, targetNode, moveType) {
56              return targetNode ? targetNode.drop !== false : true;
57          }
58          function setCheck() {
59              var zTree = $.fn.zTree.getZTreeObj("treeDemo"),
60              isCopy = $("#copy").attr("checked"),
61              isMove = $("#move").attr("checked"),
62              prev = $("#prev").attr("checked"),
63              inner = $("#inner").attr("checked"),
64              next = $("#next").attr("checked");
65              zTree.setting.edit.drag.isCopy = isCopy;
```

```
66              zTree.setting.edit.drag.isMove = isMove;
67              showCode(1, ['setting.edit.drag.isCopy = ' + isCopy, 'setting.
68      edit.drag.isMove = ' + isMove]);
69              zTree.setting.edit.drag.prev = prev;
70              zTree.setting.edit.drag.inner = inner;
71              zTree.setting.edit.drag.next = next;
72              showCode(2, ['setting.edit.drag.prev = ' + prev, 'setting.edit.
73      drag.inner = ' + inner, 'setting.edit.drag.next = ' + next]);
74          }
75          function showCode(id, str) {
76              var code = $("#code" + id);
77              code.empty();
78              for (var i=0, l=str.length; i<l; i++) {
79                  code.append("<li>" + str[i] + "</li>");
80              }
81          }
82          $(document).ready(function(){
83              $.fn.zTree.init($("#treeDemo"), setting, zNodes);
84              setCheck();
85              $("#copy").bind("change", setCheck);
86              $("#move").bind("change", setCheck);
87              $("#prev").bind("change", setCheck);
88              $("#inner").bind("change", setCheck);
89              $("#next").bind("change", setCheck);
90          });
91      </SCRIPT>
```

8.3.4　zTree 实现节点的增加、删除、修改功能

下面的示例建立了一个 zTree，并在树的节点实现增加、删除、修改的功能。

【例 8.21】建立树，并对树实现节点的增加、删除、修改功能，结果如图 8.10 所示。

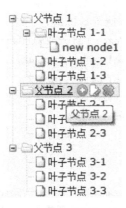

图 8.10　节点的增加、删除、修改功能

实现增加、删除、修改功能的代码如下。

```
1     <link rel="stylesheet" href="../../../css/zTreeStyle/zTreeStyle.css" type
2    ="text/css">
3     <script type="text/javascript" src="../../../js/jquery-1.4.4.min.js"></
4    script>
5     <script type="text/javascript" src="../../../js/jquery.ztree.core 3.5.js"
6    ></script>
7     <script type="text/javascript" src="../../../js/jquery.ztree.excheck
8    3.5.js"></script>
9     <script type="text/javascript" src="../../../js/jquery.ztree.exedit
10    3.5.js"></script>
11     <SCRIPT type="text/javascript">
12         <!--
13         var setting = {
14             view: {
15                 addHoverDom: addHoverDom,
16                 removeHoverDom: removeHoverDom,
17                 selectedMulti: false
18             },
19             edit: {
20                 enable: true,
21                 editNameSelectAll: true,
22                 showRemoveBtn: showRemoveBtn,
23                 showRenameBtn: showRenameBtn
24             },
25             data: {
26                 simpleData: {
27                     enable: true
28                 }
29             },
30             callback: {
31                 beforeDrag: beforeDrag,
32                 beforeEditName: beforeEditName,
33                 beforeRemove: beforeRemove,
34                 beforeRename: beforeRename,
35                 onRemove: onRemove,
36                 onRename: onRename
37             }
38         };
39         var zNodes = [
40             { id:1, pId:0, name:"父节点 1", open:true},
41             { id:11, pId:1, name:"叶子节点 1-1"},
42             { id:12, pId:1, name:"叶子节点 1-2"},
```

```
43              { id:13, pId:1, name:"叶子节点 1-3"},
44              { id:2, pId:0, name:"父节点 2", open:true},
45              { id:21, pId:2, name:"叶子节点 2-1"},
46              { id:22, pId:2, name:"叶子节点 2-2"},
47              { id:23, pId:2, name:"叶子节点 2-3"},
48              { id:3, pId:0, name:"父节点 3", open:true},
49              { id:31, pId:3, name:"叶子节点 3-1"},
50              { id:32, pId:3, name:"叶子节点 3-2"},
51              { id:33, pId:3, name:"叶子节点 3-3"}
52          ];
53          var log, className = "dark";
54          function beforeDrag(treeId, treeNodes) {
55              return false;
56          }
57          function beforeEditName(treeId, treeNode) {
58              className = (className === "dark" ? "":"dark");
59              showLog("[ " + getTime() + " beforeEditName ]    
60      " + treeNode.name);
61              var zTree = $.fn.zTree.getZTreeObj("treeDemo");
62              zTree.selectNode(treeNode);
63              return confirm("进入节点 -- " + treeNode.name + " 的编辑状态吗?");
64          }
65          function beforeRemove(treeId, treeNode) {
66              className = (className === "dark" ? "":"dark");
67              showLog("[ " + getTime() + " beforeRemove ]     " +
68      treeNode.name);
69              var zTree = $.fn.zTree.getZTreeObj("treeDemo");
70              zTree.selectNode(treeNode);
71              return confirm("确认删除节点 -- " + treeNode.name + " 吗?");
72          }
73          function onRemove(e, treeId, treeNode) {
74              showLog("[ " + getTime() + " onRemove ]     " +
75      treeNode.name);
76          }
77          function beforeRename(treeId, treeNode, newName, isCancel) {
78              className = (className === "dark" ? "":"dark");
79              showLog((isCancel ? "< span style = 'color:red' >":"") + "[ " + get
80      Time() + " beforeRename ]     " + treeNode.name +
81      (isCancel ? "</span>":""));
82              if (newName.length ==0) {
83                  alert("节点名称不能为空.");
84                  var zTree = $.fn.zTree.getZTreeObj("treeDemo");
85                  setTimeout(function(){zTree.editName(treeNode)}, 10);
```

```
86              return false;
87          }
88          return true;
89      }
90      function onRename(e, treeId, treeNode, isCancel) {
91          showLog((isCancel ? "<span style='color:red'>":"") + "[ " + get
92  Time() + " onRename ]     " + treeNode.name + (isCancel
93  ? "</span>":""));
94      }
95      function showRemoveBtn(treeId, treeNode) {
96          return !treeNode.isFirstNode;
97      }
98      function showRenameBtn(treeId, treeNode) {
99          return !treeNode.isLastNode;
100     }
101     function showLog(str) {
102         if (!log) log = $("#log");
103         log.append("<li class='" + className + "'>" + str + "</li>");
104         if(log.children("li").length>8) {
105             log.get(0).removeChild(log.children("li")[0]);
106         }
107     }
108     function getTime() {
109         var now = new Date(),
110         h = now.getHours(),
111         m = now.getMinutes(),
112         s = now.getSeconds(),
113         ms = now.getMilliseconds();
114         return (h+":"+m+":"+s+" "+ms);
115     }
116     var newCount = 1;
117     function addHoverDom(treeId, treeNode) {
118         var sObj = $("#" + treeNode.tId + "_span");
119         if (treeNode.editNameFlag || $("#addBtn_" + treeNode.tId).length>0)
120     return;
121         var addStr = "<span class='button add' id='addBtn_" + treeNode.tId
122             + "' title='add node' onfocus='this.blur();'></span>";
123         sObj.after(addStr);
124         var btn = $("#addBtn_" + treeNode.tId);
125         if (btn) btn.bind("click", function(){
126             var zTree = $.fn.zTree.getZTreeObj("treeDemo");
127             zTree.addNodes(treeNode, {id:(100 + newCount), pId:treeNode.id,
128     name:"new node" + (newCount++)});
```

```
129                        return false;
130                   });
131              };
132              function removeHoverDom(treeId, treeNode) {
133                   $("#addBtn_" + treeNode.tId).unbind().remove();
134              };
135              function selectAll() {
136                   var zTree = $.fn.zTree.getZTreeObj("treeDemo");
137                   zTree.setting.edit.editNameSelectAll = $("#selectAll").attr("checked");
138              }
139              $(document).ready(function(){
140                   $.fn.zTree.init($("#treeDemo"), setting, zNodes);
141                   $("#selectAll").bind("click", selectAll);
142              });
143              //-->
144         </SCRIPT>
```

8.4　JQGrid 表格控件

JQGrid 是 JQuery 的一种开源的表格控件包。它的界面风格有些类似于 Excel，可以实现自动分页、按要求排序、列属性冻结、设定字体颜色、设定不同行的属性、单击/双击事件等效果。针对大数据的情况，它自身有可靠的机制，可以完成对大数据的快速加载。JQGrid 的基本页面效果如图 8.11 所示。

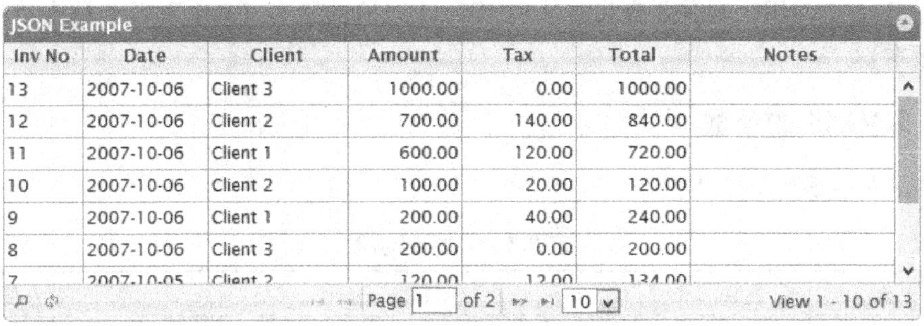

图 8.11　JQGrid 样例

8.4.1　JQGrid 的原理

JQGrid 是典型的 B/S 架构，服务器端只提供数据管理，客户端只提供数据显示。换句话说，JQGrid 可以以一种更加简单的方式来展现用户数据库的信息，而且也可以把客户端数据传回给服务器端。

对于 JQGrid，我们所关心的是：必须有一段代码把一些页面信息保存到数据库中，而且也能够把响应信息返回给客户端。JQGrid 是用 AJAX 来实现对请求与响应的处理。

8.4.2 JQGrid 的安装

JQGrid 的安装方法和 zTree 相似。两者都是基于 JQuery 和 CSS 样式来实现的工具。因而，在使用 JQGrid 时，将相应的 js 文件和 CSS 样式引入即可。需要引入的文件如下所示。

```
1    <link rel="stylesheet" type="text/css" media="screen" href="css/ui-
2    lightness/jquery-ui-1.7.1.custom.css" />
3    <link rel="stylesheet" type="text/css" media="screen" href="js/src/
4    css/ui.jqgrid.css" />
5    <link rel="stylesheet" type="text/css" media="screen" href="js/src/
6    css/jquery.searchFilter.css" />
7    <script src="js/jquery-1.3.2.min.js" type="text/javascript"></script>
8    <script src="js/i18n/grid.loader-cn.js" type="text/javascript"></script>
9    <style>
10   html, body {
11     margin: 0;
12     padding: 0;
13     font-size: 75%;
14   }
15   </style>
```

注意：由于 jquery-ui 的字体大小与 JQGrid 的字体大小是不一致的，因此需要在页面上再加上一段 style 来指定页面上文字的大小。另外，由于 grid.loader 有多种语言，这里引用的是中文版本的 JQGrid，因此引用 grid.loader-cn.js。如果要引入其他语言版本，则需要按照需求自行查找相应的 js 文件。目前 JQGrid 已经推出了 4.6.0 版本，已经不再支持 IE 6。如果需要在 IE 6 上使用，则需要查找并使用 JQGrid 以前的版本（4.4.3 版本是支持 IE 6 的最后一个版本）。

8.4.3 JQGrid 的参数

JQGrid 的参数见表 8.1。

表 8.1 JQGrid 参数

名称	类型	描述	默认值	可修改
url	string	获取数据的地址		
datatype	string	从服务器端返回的数据类型，默认为 xml。可选类型：xml、local、json、jsonnp、script、xmlstring、jsonstring、clientside		
mtype	string	AJAX 提交方式：POST 或者 GET，默认为 GET		
colNames	Array	列显示名称，是一个数组对象		
colModel	Array	常用到的属性：name，列显示的名称；index，传到服务器端用来排序的列名称；width，列宽度；align，对齐方式；sortable，是否可以排序		
pager	string	定义翻页用的导航栏，必须是有效的 HTML 元素。翻页工具栏可以放置在 HTML 页面任意位置		

（续）

名称	类型	描述	默认值	可修改
rowNum	int	在 grid 上显示记录条数，这个参数是要被传递到后台		
rowList	array	一个下拉列表框，用来改变显示记录数，当选择时会覆盖 rowNum 参数传递到后台		
sortname	string	默认的排序列。可以是列名称或者是一个数字，这个参数会被提交到后台		
viewrecords	boolean	定义是否要显示总记录数		
caption	string	表格名称		
ajaxGridOptions[a1]	object	对 AJAX 参数进行全局设置，可以覆盖 AJAX 事件	null	是
ajaxSelectOptions[a2]	object	对 AJAX 的 select 参数进行全局设置	null	是
altclass	string	用来指定行显示的 CSS，可以编辑自己的 CSS 文件，只有当 altRows 设为 ture 时才起作用	ui-priority-secondary	
altRows	boolean	设置表格 zebra-striped 值		
autoencode	boolean	对 URL 进行编码	false	是
autowidth	boolean	如果为 ture，则当表格在首次被创建时会根据父元素比例重新调整表格宽度。如果父元素宽度改变，为了使表格宽度能够自动调整，则需要实现函数：setGridWidth	false	否
cellLayout	integer	定义了单元格 padding + border 宽度。通常不必修改此值。初始值为 5	5	是
cellEdit	boolean	启用或者禁用单元格编辑功能	false	是
cellsubmit	string	定义了单元格内容保存位置	'remote'	是
cellurl	string	单元格提交的 URL	空值	是
datastr	string	xmlstring 或者 jsonstring	空值	是
deselectAfterSort	boolean	只有当 datatype 为 local 时才起作用。当排序时，不选择当前行	true	是
direction	string	表格中文字的显示方向，从左向右（ltr）或者从右向左（rtl）	ltr	否
editurl	string	定义对 form 编辑时的 URL	空值	是
emptyrecords	string	返回的数据行数为 0 时显示的信息。只有当属性 viewrecords 设置为 true 时才起作用		是
expandColClick	boolean	当其为 true，点击展开行的文本时，treeGrid 就能展开或者收缩，不仅仅是点击图片	true	否
expandColumn	string	指定哪一列来展开 treeGrid，默认为第一列，只有在 treeGrid 为 true 时才起作用	空值	否
footerrow[a3]	boolean	当其为 true 时，会在翻页栏之上增加一行	false	否

（续）

名称	类型	描述	默认值	可修改
forceFit	boolean	当其为 ture 时，调整列宽度不会改变表格的宽度。当 shrinkToFit 为 false 时，此属性会被忽略	false	否
gridstate	string	定义当前表格的状态：visible 或 hidden	visible	否
gridview	boolean	构造一行数据后添加到 grid 中，如果设为 true，则是将整个表格的数据都构造完成后再添加到 grid 中，但 treeGrid、subGrid 和 afterInsertRow 不能用	false	是
height	mixed	表格高度，可以是数字、像素值或者百分比	150	否
hiddengrid	boolean	当其为 true 时，表格不会显示，而只显示表格的标题。只有当单击显示表格的那个按钮时才会去初始化表格数据	false	否
hidegrid	boolean	启用或者禁用控制表格显示、隐藏的按钮，只有当 caption 属性不为空时才起作用	true	否
hoverrows	boolean	当其为 false 时，mouse hovering 会被禁用	false	是
jsonReader	array	描述 JSON 数据格式的数组		否
lastpage	integer	只读属性，定义了总页数	0	否
lastsort	integer	只读属性，定义了最后排序列的索引，从 0 开始	0	否
loadonce	boolean	如果其为 true，则数据只从服务器端抓取一次，之后所有操作都是在客户端执行，翻页功能会被禁用	false	否
loadtext	string	请求或者排序时显示的文字内容	Loading....	否
loadui	string	表示当执行 AJAX 请求时要做什么。disable：禁用 AJAX 执行提示；enable：默认值，执行 AJAX 请求时的提示；block：启用 Loading 提示，但阻止其他操作	enable	是
multikey	string	只有当 multiselect 设置为 true 时才起作用，定义使用哪个 key 来做多选，包括 shiftKey、altKey、ctrlKey	空值	是
multiboxonly	boolean	只有当 multiselect = true 时才起作用。当 multiboxonly 为 true 时，只有选择 checkbox 才会起作用	false	是
multiselect	boolean	定义是否可以多选	false	否
multiselectWidth	integer	当 multiselect 为 true 时，设置 multiselect 列宽度	20	否
page	integer	设置初始的页码	1	是
pagerpos	string	指定分页栏的位置	center	否
pgbuttons	boolean	是否显示翻页按钮	true	否
pginput	boolean	是否显示跳转页面的输入框	true	否
pgtext	string	当前页信息		是
prmNames	array	Default valuesprmNames：{page:"page",rows:"rows", sort:" sidx",order: "sord", search:"_ search", nd:"nd", npage:null} 当参数为 null 时，不会被发送到服务器端	空 array	是

（续）

名称	类型	描述	默认值	可修改
postData	array	此数组内容直接赋值到 url 上，参数类型：{name1：value1...}	空 array	是
reccount	integer	只读属性，定义了 grid 中确切的行数。通常情况下与 records 属性相同，但有一种情况例外，假如 rowNum = 15，但是从服务器端返回的记录数为 20，那么 records 值是 20，但 reccount 值仍然为 15，而且表格中也只显示 15 条记录	0	否
recordpos	string	定义了记录信息的位置：left、center、right	right	否
records	integer	只读属性，定义了返回的记录数	0	否
recordtext	string	显示记录数信息。{0} 为记录数开始，{1} 为记录数结束。viewrecords 为 ture 时才能起作用，且总记录数大于 0 时才会显示此信息		
resizeclass	string	定义一个 class 到一个列上用来显示列宽度调整时的效果	空值	否
rowList	array	一个数组用来调整表格显示的记录数，此参数值会替代 rowNum 参数值传给服务器端	空 array	否
rownumbers	boolean	如果其为 true，则会在表格左边新增一列，显示行顺序号，从 1 开始递增。此列名为 "rn"	false	否
rownumWidth	integer	如果 rownumbers 为 true，则可以设置 column 的宽度	25	否
savedRow	array	只读属性，只用于在编辑模式下保存数据	空值	否
scroll	boolean	创建一个动态滚动的表格。当其为 true 时，翻页栏被禁用，使用垂直滚动条加载数据，且在首次访问服务器端时将加载所有数据到客户端。当此参数为数字时，表格只控制可见的几行，所有数据都在这几行中加载	false	否
scrollOffset	integer	设置垂直滚动条宽度	18	否
scrollrows	boolean	当其为 true 时，让所选择的行可见	false	是
selarrrow	array	只读属性，用来存放当前选择的行	array	否
selrow	string	只读属性，最后选择行的 id	null	否
shrinkToFit	boolean	此属性用来说明初始化列宽度时的计算类型。如果其为 true，则按比例初始化列宽度；如果其为 false，则列宽度使用 colModel 指定的宽度	true	否
sortable	boolean	是否可排序	false	否
sortname	string	排序列的名称，此参数会被传递到后台	空字符串	是
sortorder	string	排序类型，即升序或者降序（asc 或 desc）	asc	是
subGrid	boolean	是否使用 subgrid	false	否
subGridModel	array	subgrid 模型	array	否
subGridType	mixed	如果为空，则使用表格的 dataType	null	是
subGridUrl	string	加载 subgrid 数据的 url，JQGrid 会把每行的 id 值加到 url 中	空值	是

（续）

名　称	类型	描　述	默认值	可修改
subGridWidth	integer	subgrid 列的宽度	20	否
toolbar	array	表格的工具栏。数组中有两个值，第一个为是否启用，第二个指定工具栏位置（相对于 body layer），如：[true, "both"]。工具栏位置可选值：top、bottom、both。如果工具栏在上面，则工具栏 id 为"t_"+表格 id；如果在下面，则为"tb_"+表格 id；如果只有一个工具栏，则为"t_"+表格 id	[false,""]	否
totaltime	integer	只读属性，计算加载数据的时间。目前支持 xml 和 Json 数据	0	否
treedatatype	mixed	数据类型，通常情况下与 datatype 相同，不会改变	null	否
treeGrid	boolean	启用或者禁用 treeGrid 模式	false	否
treeGridModel	string	treeGrid 所使用的方法	Nested	否
treeIcons	array	树的图标，默认值：{plus:'ui-icon-triangle-1-e',minus:'ui-icon-triangle-1-s',leaf:'ui-icon-radio-off'}		否
treeReader	array	扩展表格的 colModel 且加在 colModel 定义的后面		否
tree_root_level	numeric	root 元素的级别	0	否
userData	array	从 request 中取得的一些用户信息	array	否
userDataOnFooter	boolean	当其为 true 时，把 userData 放到底部。用法：如果 userData 的值与 colModel 的值相同，那么此列就显示正确的值，如果不相同，那么此列就为空	false	是
viewrecords	boolean	是否要显示总记录数	false	否
viewsortcols	array	定义排序列的外观和行为。数据格式：[false,'vertical',true]。第一个参数表示是否都要显示排序列的图标，false 就是只显示当前排序列的图标；第二个参数是指图标如何显示，vertical：排序图标垂直放置，horizontal：排序图标水平放置；第三个参数指单击功能，true：单击列可排序，false：单击图标排序。说明：如果第三个参数为 false，则第一个参数必须为 true，否则不能排序		否
width	number	如果 width 已经设置，则按此设置值进行操作，如果没有设置，则按 colModel 中定义的宽度计算	0	否
xmlReader	array	对 XML 数据结构的描述		否

8.4.4　JQGrid 中 ColModel 的 API

ColModel 是 JQGrid 中比较重要的一个参数，用来设置表格列的属性，见表 8.2。

表 8.2 ColModel 的属性

属性	数据类型	说明	默认值
align	string	left、center、right	left
classes	string	设置列的 CSS。多个 class 之间用空格分隔，如：'class1 class2'。表格默认的 css 属性是 ui-ellipsis	empty string
datefmt	string	"/"、"-" 和 "." 都是有效的日期分隔符。y、Y；年；YY、yy 月；m、mm 日；d、dd 日	ISO Date (Y-m-d)
defval	string	查询字段的默认值	空
editable	boolean	单元格是否可编辑	false
editoptions	array	编辑的一系列选项。{name:'_department_id',index:'_department_id',width:200,editable:true,edittype:'select',editoptions:{dataUrl:"${ctx}/admin/deplistforstu.action"}}，这个是演示从服务器端动态获取数据	empty
editrules	array	编辑的规则。例如，{name:'age',index:'age',width:90,editable:true,editrules:{edithidden:true,required:true,number:true,minValue:10,maxValue:100}}，设定年龄的最大值为100，最小值为10，而且为数字类型，并且为必须输入字段	empty
edittype	string	可以编辑的类型。可选值：text、textarea、select、checkbox、password、button、image 和 file	text
fixed	boolean	列宽度是否要固定不可变	false
formoptions	array	对 form 进行编辑时的属性设置	empty
formatoptions	array	对某些列进行格式化时的设置	none

formatter	mixed	对列进行格式化时设置的函数名或者类型。例如，{name:'sex',index:'sex',align:'center',width:60,editable:true,edittype:'select',editoptions:{value:'0:待定;1:男;2:女'},formatter:function(cellvalue,options,rowObject){
		var temp = "<img src='${ctx}/jquery-ui-1.7.2.custom/css/img/"
		if(cellvalue==1){
		temp = temp + "user-white.png";
		} else if(cellvalue==2){
		temp = temp + "user-white-female.png";
		} else {
		temp = temp + "user-silhouette.png";
		}
		temp = temp + " 'border='0'/>"
		return temp;
		}},//返回代表性别的图标
			none

hidedlg	boolean	是否显示或者隐藏此列	false
hidden	boolean	在初始化表格时是否要隐藏此列	false
index	string	索引。其和后台交互的参数为 sidx	empty

（续）

属性	数据类型	说 明	默认值
jsonmap	string	定义了返回的 JSON 数据映射	none
key	boolean	当从服务器端返回的数据中没有 id 时，将此作为唯一 rowid 使用，只有一个列可以进行这项设置。如果设置多于一个，那么只选取第一个，其他被忽略	false
label	string	如果 colNames 为空，则用此值来作为列的显示名称，如果都没有设置，则使用 name 值	none
name	string	表格列的名称，所有关键字、保留字都不能作为名称使用，包括 subgrid、cb 和 rn	Required
resizable	boolean	是否可以被 resizable	true
search	boolean	在搜索模式下，定义此列是否可以作为搜索列	true
searchoptions	array	设置搜索参数	empty
sortable	boolean	是否可排序	true
sorttype	string	当 datatype 为 local 时，定义搜索列的类型，可选值：int/integer，对 integer 排序；float/number/currency，排序数字；date，排序日期 text，排序文本	text
stype	string	定义搜索元素的类型	text
surl	string	搜索数据时的 URL	empty
width	number	默认列的宽度，只能是像素值，不能是百分比	150
xmlmap	string	定义当前列和返回的 XML 数据之间的映射关系	none
unformat	function	unformat 单元格值	null

8.4.5 JQGrid 的代码格式

JQGrid 的代码格式并不复杂，如下所示。

HTML 代码示例如下。

```
1    <table id="list2"></table>
2    <div id="pager2"></div>
```

Java Script 代码示例如下。

```
1    jQuery("#list2").jqGrid({
2        url:'server.php? q=2',
3        datatype: "json",
4        colNames:['Inv No','Date','Client','Amount','Tax','Total','Notes'],
5        colModel:[
6            {name:'id',index:'id', width:55},
7            {name:'invdate',index:'invdate', width:90},
8            {name:'name',index:'name asc, invdate', width:100},
9            {name:'amount',index:'amount', width:80, align:"right"},
10           {name:'tax',index:'tax', width:80, align:"right"},
```

```
11              {name:'total',index:'total', width:80,align:"right"},
12              {name:'note',index:'note', width:150, sortable:false}
13          ],
14          rowNum:10,
15          rowList:[10,20,30],
16          pager: '#pager2',
17          sortname: 'id',
18          viewrecords: true,
19          sortorder: "desc",
20          caption:"JSON Example"
21      });
22      jQuery("#list2").jqGrid('navGrid','#pager2',{edit:false,add:false,del:
23  false});
```

在上述代码中，url 是访问后台的数据的地址；datatype 是返回值的数据类型；colNames 是列的名称；colModel 定义对应列的属性（和 colNames 对应，前后顺序进行对照）；rowNum 用于定义初始分页每页的行数；rowList 定义 table 所有的行数的选项；pager 用于定义分页栏。

8.5　JQGrid 实例

【例 8.22】建立基本的 JQGrid。

添加一个列名分别为"产品 ID"、"名称"、"产品数量"、"颜色"、"安全库存"、"再订购点"、"标准成本"、"原价"、"guid"的 JQGrid 表格。

笔者使用的基本是通过 JSON 来传递数据的方法，代码如下所示。

页面代码：

```
1   $ (document).ready(function() {
2       jQuery("#list").jqGrid({
3           url: 'data/load.aspx',
4           datatype: "json",
5           height: 'auto',
6           colNames: ['产品 ID','名称','产品数量','颜色','安全库存','再订购点','
7   标准成本','原价','guid'],
9           colModel: [
10                      { name: 'ProductID', index: 'ProductID', sortable: false,
11  width: 100 },
12                      { name: 'Name', index: 'Name', sorttype: "string", width: 180 },
13                      { name: 'ProductNumber', index: 'ProductNumber', sorttype: "
14  string", width: 100 },
15                      { name: 'Color', index: 'Color', sorttype: "string", ed
16  itable: true, width: 60 },
17                      { name: 'SafetyStockLevel', index: 'SafetyStockLevel', a-
18  lign: "right", sorttype: "float", width: 100 },
```

```
19                    { name: 'ReorderPoint', index: 'ReorderPoint', align:
20     "right",sorttype: "float", width: 100 },
21                    { name: 'StandardCost', index: 'StandardCost', align:
22     "right", sorttype: "float", width: 100 },
23                    { name: 'ListPrice', index: 'ListPrice', align: "right",
24     sorttype: "float", width: 100 },
25                    { name: 'rowguid', index: 'rowguid', sorttype: "string",
26     editable: true, width: 250 }
27                  ],
28          viewrecords: true,
29          rowNum: 10,
30          rowList: [10, 20, 30, 40, 50, 100, 500],
31          sortname: 'id',
32          jsonReader: {
33              root: "griddata",
34              total: "totalpages",
35              page: "currpage",
36              records: "totalrecords",
37              repeatitems: false
38          },
39          pager: "#pager",
40          caption: "产品列表",
41          sortorder: "desc",
42          hidegrid: false
43       });
44       jQuery("#list").jqGrid('navGrid', '#pager', { edit: false, add: false,
45    del: false });
46    });
```

JQGrid 代码（index.js）：

```
1     $(document).ready(function () {
2        jQuery("#list").jqGrid({
3           url: 'data/load.aspx',
4           datatype: "json",
5           height: 'auto',
6           colNames: ['产品ID', '名称', '产品数量', '颜色', '安全库存', '再订购点', '
7    标准成本', '原价', 'guid'],
8           colModel: [
9                    { name: 'ProductID', index: 'ProductID', sortable: false,
10    width: 100 },
11                   { name: 'Name', index: 'Name', sorttype: "string",
12    width: 180 },
13                   { name: 'ProductNumber', index: 'ProductNumber', sorttype:
14    "string", width: 100 },
```

```
15                       { name: 'Color', index: 'Color', sorttype: "string",
16       editable: true, width: 60 },
17                       { name: 'SafetyStockLevel', index: 'SafetyStockLevel',
18       align: "right", sorttype: "float", width: 100 },
19                       { name: 'ReorderPoint', index: 'ReorderPoint', align:
20       "right", sorttype: "float", width: 100 },
21                       { name: 'StandardCost', index: 'StandardCost', align:
22       "right", sorttype: "float", width: 100 },
23                       { name: 'ListPrice', index: 'ListPrice', align: "right",
24       sorttype: "float", width: 100 },
25                       { name: 'rowguid', index: 'rowguid', sorttype: "string",
26       editable: true, width: 250 }
27                   ],
28           viewrecords: true,
29           rowNum: 10,
30           rowList: [10, 20, 30, 40, 50, 100, 500],
31           sortname: 'id',
32           jsonReader: {
33               root: "griddata",
34               total: "totalpages",
35               page: "currpage",
36               records: "totalrecords",
37               repeatitems: false
38           },
39           pager: "#pager",
40           caption: "产品列表",
41           sortorder: "desc",
42           hidegrid: false
43       });
44       jQuery("#list").jqGrid('navGrid', '#pager', { edit: false, add: false,
45       del: false });
46   });
```

后台代码为：

```
1    using System;
2    using System.Collections.Generic;
3    using System.Web;
4    using System.Web.UI;
5    using System.Web.UI.WebControls;
6    using System.Data;
7    using System.Text;
9    public partial class jqGrid_data_load : System.Web.UI.Page
10   {
11       protected void Page_Load(object sender, EventArgs e)
```

```
12        {
13            string _search = Request.Params["_search"];
14            string sidx = Request.Params["sidx"];
15            string sord = Request.Params["sord"];
16            string searchOper = Request.Params["searchOper"];
17            string searchString = Request.Params["searchString"];
18            string searchField = Request.Params["searchField"];
19            string filters = Request.Params["filters"];
20            string page = Request.Params["page"];//当前页面
21            string rows = Request.Params["rows"];
22            string sql = "select top " + (int.Parse(page) * int.Parse(rows)) + "
23   * from Production.Product " +"where ProductID not in (" +"select top " +
24   ((int.Parse(page)-1) * int.Parse(rows)) + " ProductID from Produc
25   tion.Product order by ProductID asc)" +" order by ProductID asc";
26            if (_search == "true")
27            {
28                string where = searchField + " = '" + searchString + "'";
29            }
30            DataTable dt = ZQ.DbHelper.GetDataTable(sql);
31            string total = ZQ.DbHelper.GetSingle("select count(*) from Produc
32   tion.Product").ToString();
33            StringBuilder sb = new StringBuilder();
34   sb.Append(" {\"totalpages\":\"" + Convert.ToInt32(int.Parse(total)/
35   int.Parse(rows)) + "\",\"currpage\":\"" + page + "\",\"totalrecords\":\"" + to
36   tal + "\",\"griddata\":[");
37            foreach (DataRow row in dt.Rows)
38            {
39                sb.Append("{");
40                foreach (DataColumn col in dt.Columns)
41                {
42                    sb.Append("\"" + col.ColumnName + "\":" + "\"" + row
43   [col.ColumnName] + "\",");
44                }
45                sb.Remove(sb.Length-1,1);
46                sb.Append("},");
47            }
48            sb.Remove(sb.Length-1, 1);
49            sb.Append("]}");
50            Response.Write(sb.ToString());
51        }
52        public string getSymblo(string s, string value, string field)
53        {
54            string symbol = "";
```

```
55          switch (s)
56          {
57              case "eq":
58                  symbol = " = ";
59                  break;
60              case "ne":
61                  symbol = "! = ";
62                  break;
63              case "lt":
64                  symbol = " < ";
65                  break;
66              case "le":
67                  symbol = " < = ";
68                  break;
69              case "gt":
70                  symbol = " > ";
71                  break;
72              case "ge":
73                  symbol = " > = ";
74                  break;
75              case "bw":
76                  symbol = "LIKE";
77                  break;
78              case "bn":
79                  symbol = "NOT LIKE";
80                  break;
81              case "in":
82                  symbol = "";
83                  break;
84              default:
85                  break;
86          }
87          return symbol;
88      }
89  }
```

【例 8.23】冻结 JQGrid 的列。

这里提到的冻结的作用与 Excel 中的冻结是一样的，是使从开头开始某几列不会随着表格内容的移动而移动。基于例 8.22 的表格，我们冻结前两列，需要修改和添加的代码为：

```
1   colModel: [
2           { name: 'ProductID', index: 'ProductID', sortable: false, width: 100,
3       frozen: true },
4           { name: 'Name', index: 'Name', sorttype: "string", width: 180, frozen:
5       true },
```

```
6              { name: 'ProductNumber', index: 'ProductNumber', sorttype: "string",
7       width: 100 },
8              { name: 'Color',index: 'Color', sorttype: "string", editable: true,
9       width: 60 },
10             { name: ' SafetyStockLevel ', index: ' SafetyStockLevel ', align:
11     "right", sorttype: "float", width: 100 },
12             { name: 'ReorderPoint', index: 'ReorderPoint', align: "right", sort-
13     type: "float", width: 100 },
14             { name: 'StandardCost', index: 'StandardCost', align: "right", sort-
15     type: "float", width: 100 },
16             { name: 'ListPrice', index: 'ListPrice', align: "right", sorttype: "
17     float", width: 100 },
18             { name: 'rowguid', index: 'rowguid', sorttype: "string", editable:
19     true, width: 250 }
20          ],
21     jQuery("#list").jqGrid('setFrozenColumns');
```

【例 8.24】 JQGrid 二级表头。

如 Excel 一样，对 JQGrid 添加二级表头（到 v4.4.3 为止，只能添加到二级表头，不能添加更多的，v4.5 以上的版本支持三级表头）。

基于例 8.22，对"安全库存"，"再订购点"进行合并，合并后名称为"二次供需"，将"标准成本"、"原价"合并为"价格属性"，添加代码如下所示。

```
1       $ ("#list").jqGrid('setGroupHeaders', {
2      useColSpanStyle : true, //没有表头的列是否与表头列位置的空单元格合并
3      groupHeaders : [ {
4      startColumnName : 'SafetyStockLevel', //对应 colModel 中的 name
5      numberOfColumns : 2, //跨越的列数
6      titleText : '二次供需'
7      },
8      {
9      startColumnName : 'StandardCost', //对应 colModel 中的 name
10     numberOfColumns : 2, //跨越的列数
11     titleText : '价格属性'
12     }]
13     });
```

【例 8.25】 JQGrid 分页。

实现 JQGrid 自动分页功能。使用 JQGrid 自带分页功能，并将其分为一页 10、20、30 个文件。实现代码如下所示。

前台代码为：

```
1       <! DOCTYPE html PUBLIC "-//W3C//DTD XHTML 1.0 Transitional//EN" "http://
2       www.w3.org/TR/xhtml1/DTD/xhtml1-transitional.dtd" >
3       <html xmlns = "http://www.w3.org/1999/xhtml" >
4       <head runat = "server" >
```

```
5       <title>评论管理</title>
6       <!--引入主题文件-->
7       <link rel="stylesheet" type="text/css" media="screen" href="../
8       themes/redmond/jquery-ui-1.8.4.custom.css" />
9       <link rel="stylesheet" type="text/css" media="screen" href="../
10      themes/ui.jqgrid.css" />
11      <!--引入脚本文件-->
12      <script type="text/javascript" src="../scripts/jQuery/jquery-
13      1.4.2.min.js"></script>
14      <script type="text/javascript" src="../scripts/jQuery/plugins/
15      jquery-ui-1.8.4.custom.min.js"></script>
16      <script type="text/javascript" src="../scripts/jQuery/plugins/
17      grid.locale-cn.js"></script>
18      <script type="text/javascript" src="../scripts/jQuery/plugins/
19      jquery.jqGrid.min.js"></script>
20      <script type="text/javascript">
21          $(document).ready(function () {
22              jQuery("#list").jqGrid({
23                  url: 'asynchronous/GridData.ashx?p=Comment',
24                  datatype: "json",
25                  height: 'auto',
26                  colNames: ['评论ID', '类别ID', '文章ID', '留言人', '留言内容', '
27      发布日期', '留言IP'],
28                  colModel: [
29                      { name: 'CommentID', index: 'CommentID', sorttype: "int",
30      width: 60 },
31                      { name: 'TypeID', index: 'TypeID', sorttype: "int", width: 60 },
32                      { name: 'FromID', index: 'FromID', sorttype: "int", width: 60 },
33                      { name: 'Name', index: 'Name', editable: true, width: 60 },
34                      { name: 'Contents', index: 'Contents', sortable: false,
35      width: 300 },
36                      { name: 'PublishDate', index: 'PublishDate', sorttype: "
37      date", width: 190 },
38                      { name: 'IP', index: 'IP', align: "right", sorttype: "
39      float", editable: true, width: 130 }
40                  ],
41                  viewrecords: true,
42                  rowNum: 10,
43                  rowList: [10, 20, 30],
44                  sortname: 'CommentID',
45                  jsonReader: {
46                      root: "griddata",
47                      total: "totalpages",
```

```
48                        page: "currpage",
49                        records: "totalrecords",
50                        repeatitems: false
51                    },
52                    pager: "#pager",
53                    caption: "评论管理",
54                    sortorder: "desc",
55                    hidegrid: false
56                });
57                jQuery("#list").jqGrid('navGrid', '#pager', { edit: false, add:
58       false, del: false });
59            });
60        </script>
61    </head>
62    <body>
63        <form id="frmMComment" runat="server">
64        <div>
65            <table id="list" class="scroll" cellpadding="0" cellspacing="0"></table><!
66    --用于数据显示-->
67        </table>
68            <div id="pager" class="scroll" style="text-align: center;"><!--用
69    于分页的层-->
70        </div>
71        </div>
72        </form>
73    </body>
74    </html>
```

后台代码为：

```
1     <%@ WebHandler Language="C#" Class="GridData" %>
2     using System;
3     using System.Web;
4     using System.Data;
5     using Wood8.Common;
6     using Wood8.DataAccess.SQLServer;
7     using Newtonsoft.Json;
8     using Newtonsoft.Json.Converters;
9     using System.Web.Services;
10    using System.Collections;
11    using System.Collections.Generic;
12    [WebService(Namespace = "http://tempuri.org/")]
13    [WebServiceBinding(ConformsTo = WsiProfiles.BasicProfile1_1)]
14    public class GridData : IHttpHandler {
15        public void ProcessRequest(HttpContext context)
```

```csharp
16          {
17              context.Response.Buffer = true;
18              context.Response.ExpiresAbsolute = DateTime.Now.AddDays(-1);
19              context.Response.AddHeader("pragma", "no-cache");
20              context.Response.AddHeader("cache-control", "");
21              context.Response.CacheControl = "no-cache";
22              context.Response.ContentType = "text/plain";
23              DataTable dt;
24              string jsonData = string.Empty;
25              string sPage = HttpContext.Current.Request.Params["page"].ToString();
26  //当前请求第几页
27              int iPage = int.Parse(sPage);
28              string sLimit = HttpContext.Current.Request.Params["rows"].ToString();
29  //grid需要显示几行
30              int iLimit = int.Parse(sLimit);
31              string sSidx = HttpContext.Current.Request.Params["sidx"].ToString();
32  //按什么排序
33              string sSord = HttpContext.Current.Request.Params["sord"].ToString();
34  //排序方式(desc/asc)
35              if(sSidx == "")
36              {
37                  sSidx = "1";
38              }
39              int iTotalpages;
40              SQLComment sc = new SQLComment();
41              DataSet sResult = sc.getAllComments();
42              int iCount = sResult.Tables[0].Rows.Count;
43              if( iCount > 0 )
44              {
45                  int iR = iCount/iLimit;
46                  iTotalpages = iR + 1;
47              }
48              else
49              {
50                  iTotalpages = 0;
51              }
52              if (iPage > iTotalpages)
53              {
54                  iPage = iTotalpages;
55              }
56              int iStart = iLimit * iPage-iLimit + 1;
57              iLimit = iLimit * iPage;
58              sResult = sc.getCommentsFromTo(iStart, iLimit, sSidx, sSord, iPage);
```

```csharp
59              dt = sResult.Tables[0];
60              string totalpages = iTotalpages.ToString();
61              string currpage = iPage.ToString();
62              string totalrecords = iCount.ToString();
63              IsoDateTimeConverter idtc = new IsoDateTimeConverter();
64              idtc.DateTimeFormat = "yyyy-MM-dd hh:mm:ss ffffff";
65              jsonData = JsonConvert.SerializeObject(dt, idtc).ToString();
66              string returnData = string.Empty;
67              returnData = "{";
68              //总共多少页
69              returnData += "\"totalpages\"";
70              returnData += ":";
71              returnData += "\"";
72              returnData += totalpages;
73              returnData += "\"";
74              returnData += ",";
75              //当前第几页
76              returnData += "\"currpage\"";
77              returnData += ":";
78              returnData += "\"";
79              returnData += currpage;
80              returnData += "\"";
81              returnData += ",";
82              //总共多少记录
83              returnData += "\"totalrecords\"";
84              returnData += ":";
85              returnData += "\"";
86              returnData += totalrecords;
87              returnData += "\"";
88              returnData += ",";
89              //datable转换得到的JSON字符串
90              returnData += "\"griddata\"";
91              returnData += ":";
92              returnData += jsonData;
93              returnData += "}";
94              context.Response.Write(returnData);
95          }
96          public bool IsReusable {
97              get {
98                  return false;
99              }
100         }
101     }
```

※ 习 题

1. 使用 zTree 建立一棵树，要求：
1）有 5 个根节点，其中前 3 个根节点有下一级节点，第 4 个根节点没有下一级节点，但类型为 Parent，第 5 个根节点为叶子节点。
2）不显示连接线。
3）根节点和中间节点至少使用 3 种图标样式来作为其前面的图标。
4）叶子节点不显示图标。

2. 建立一棵 zTree 树，全部节点使用异步加载方式，并在双击节点时 alert 出节点的名称（双击时要避免节点收缩再展开）。

3. 建立一棵带有 CheckBox 的树，要求单击后面的确定按钮后 alert 出所有被选中的节点的名称。

4. 本地建立一个表格，列名为："序号"、"姓名"、"年龄"、"学号"、"班级"。初始化时添加 3 条数据到表格中，然后添加一个"添加"按钮，可以手动向表格中添加数据（不经过数据库）。

5. 将习题 4 的表格改为从后台数据库中自动加载的格式，并实现自动分页，分页按照每页 20、30、40、50 的格式来实现。

6. 对于习题 5 的表格，进行下列操作：
1）添加 Grid 默认的行号（通过 JQGrid 自身属性来完成）。
2）冻结"序号"和"姓名"两列。
3）序号、姓名、年龄和班级居中显示，学号居右显示。

7. 对于习题 5 的表格，添加 3 种宽度限制，单击相应的宽度值将表格修改为对应的宽度。

※ 综合应用

1. 使用 zTree 一次性建立 3 棵树。前两棵树只有叶子节点，且父节点和叶子节点的图标都不相同，第三棵树不使用自定义图标，使用默认的图标，结果如图 8.12 所示。

图 8.12　使用 zTree 建立 3 棵树

2. 利用JQGrid建立表格,显示访问者的名称、开始时间、终止时间、是否超时和删除按钮,并可以通过开始时间和终止时间来查询指定时间段的项目。其中,时间的输入使用日期控件来实现。结果如图8.13所示。

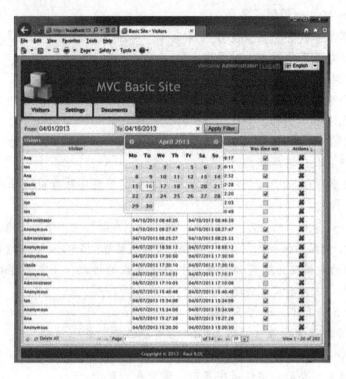

图8.13 利用JQGrid建立表格

第 9 章

AJAX 技术

AJAX 全称为 Asynchronous JavaScript And XML（异步 JavaScript 及 XML），它并非缩写词，而是由 Jesse James Gaiiett 创造的名词，是指一种创建交互式网页应用的网页开发技术。

9.1 AJAX 概述

AJAX 不是一种新的编程语言，而是一种基于现有标准的用于创建快速动态网页的技术。简单地说，AJAX 会在不刷新整个页面的前提下，对页面中的部分内容进行更新，以增强页面的交互效果，优化用户的体验。

AJAX 在 1998 年前后得到了应用，由 Outlook Web Access 小组完成了第一个允许客户端脚本发送 HTTP 请求（XMLHTTP）的组件。该组件属于微软公司的 Exchange Server，并且迅速地成为了 Internet Explorer 4.0 的一部分。

2005 年初，许多事件使得 AJAX 被大众所接受。Google 公司在它著名的交互应用程序中使用了异步通信，如 Google 讨论组、Google 地图、Google 搜索建议、Gmail 等。AJAX 这个词由《AJAX: A New Approach to Web Applications》一文（Garrett, 2005）所创造，该文的迅速流传提高了人们使用该项技术的意识。

AJAX 前景非常乐观，因为它可以提高系统性能，优化用户界面。AJAX 使用 XHTML + CSS 来表示信息，使用 JavaScript 操作 DOM（Document Object Model）进行动态显示及交互，使用 XML 和 XSLT 进行数据交换及相关操作，使用 XMLHttpRequest 对象与 Web 服务器进行异步数据交换，使用 JavaScript 将所有的东西绑定在一起。

传统的 Web 应用允许用户端填写表单（form），当提交表单时就向 Web 服务器发送一个请求。服务器接收并处理传来的表单，然后返回一个新的网页。这种做法浪费了许多带宽，因为在前后两个页面中的大部分 HTML 代码往往是相同的。由于每次应用的交互都需要向服务器发送请求，应用的响应时间就依赖于服务器的响应时间，因此，导致了用户界面的响应比本地应用慢得多。

与此不同，AJAX 应用可以仅向服务器发送并取回必需的数据，它使用 SOAP 或其他一些基于 XML 的页面服务接口，并在客户端采用 JavaScript 处理来自服务器的响应。因为在服务器和浏览器之间交换的数据大量减少（大约只有原来的 5%），所以就能看到响应更快的应用。同时，很多的处理工作可以在发出请求的客户端机器上完成，Web 服务器的处理时间也减少了。

9.2 原理简介

AJAX 是基于 JavaScript 中的 XMLHttpRequest 对象实现的。目前，主流的一些浏览器均支持 XMLHttpRequest 对象（W3School，2014）。本节将介绍如何通过 XMLHttpRequest 对象来实现 AJAX。

9.2.1 创建对象

在实现 AJAX 时，需要先创建一个 XMLHttpRequest 对象实例，如下面的代码所示。

```
var 变量名 = new XMLHttpRequest();
```

如果实际使用的浏览器版本较老（如 IE 5、IE 6 等），则需要先检查浏览器是否支持 XMLHttpRequest 对象。如果支持，则创建 XMLHttpRequest 对象；如果不支持，则创建 ActiveXObject。可以通过下面的代码来实现智能选择。

```
1    var XMR;
2    if (window.XMLHttpRequest)
3    {// IE7+、Firefox、Chrome、Opera、Safari
4        XMR = new XMLHttpRequest();
5    }
6    else
7    {// IE 6 和 IE 5
8        XMR = new ActiveXObject("Microsoft.XMLHTTP");
9    }
```

9.2.2 发送请求

在创建 XMLHttpRequest 对象实例后，需要用它向服务器发送请求，将有关数据提交给服务器。可以通过 XMLHttpRequest 对象的 open() 方法和 send() 方法来完成该过程。

（1）open（method，url，async）

作用：设定请求的类型、URL 地址、是否属于异步处理请求。

相关参数介绍如下。

method：只能是 Get 或者 Post。一般情况下，选择 Get，因为更简单也更快；但是当需要传送大量数据的时候，需要用 Post 方式。此外，Post 方法比 Get 方法更加稳定与可靠。

url：文件路径。可以是任何类型的文件，如 .txt 和 .xml；也可以是服务器脚本文件，如 .asp 和 .html。在 MVC 4 中，文件路径也可以是路由地址，即"/{Controller}/{Action}/参数"形式。

async：只能是 true（异步）或者 false（同步）。当该参数为 true 时，JavaScript 可以在等待响应的同时执行其他脚本，并在响应就绪后对响应进行处理。

（2）send（string）

作用：将请求发送到服务器。

相关参数介绍如下。

string：仅用于 Post 请求，用于设置传递的数据。

下面通过几个示例来说明 XMLHttpRequest 对象发送请求的过程。

【例 9.1】 基本的 Get 方式。

S1.cshtml 中的代码：

```
1   @{
2       Layout = null;
3   }
4   <!DOCTYPE html>
5   <html>
6   <head>
7       <meta name="viewport" content="width=device-width" />
8       <title>S1</title>
9       <script type="text/javascript">
10          function GetTime() {
11              var xmlhttp = new XMLHttpRequest();
12              xmlhttp.open("Get", "/Home/GetTime", true);
13              xmlhttp.send();
14              xmlhttp.onreadystatechange = function () {
15                  if (xmlhttp.readyState == 4 && xmlhttp.status == 200) {
16                      document.getElementById("Time").innerHTML =
17  xmlhttp.responseText;
18                  }
19              }
20          }
21      </script>
22  </head>
23  <body>
24      <button type="button" onclick="GetTime()">获取时间</button>
25      <div id="Time"></div>
26  </body>
27  </html>
```

HomeController.cs 中的有关 Action：

```
1   public ActionResult S1()
2   {
3       return View();
4   }
5   public ActionResult GetTime()
6   {
7       return Content(DateTime.Now.ToString());
8   }
```

页面的运行结果如图 9.1 所示。单击"获取时间"按钮后，将在按钮下面出现如图 9.2 所示的时间。

图 9.1　页面运行结果

图 9.2　获取的时间

注意：上述代码中的 xmlhttp.onreadystatechange 部分是获取响应和页面处理，将在 9.2.3 节和 9.2.4 节中介绍。

【例 9.2】 通过 Get 方式传递参数。

S2.cshtml 中的代码：

```
1       @{
2           Layout = null;
3       }
4       <!DOCTYPE html>
5       <html>
6       <head>
7       <meta name="viewport" content="width=device-width" />
8       <title>S2</title>
9       <script type="text/javascript">
10          function GetNumber() {
11              var xmlhttp = new XMLHttpRequest();
12              var maxValue = document.getElementById("maxValue").value;
13              var URL = "/Home/GetNumber? maxValue=" + maxValue;
14              xmlhttp.open("Get", URL, true);
15              xmlhttp.send();
16              xmlhttp.onreadystatechange = function() {
17                  if (xmlhttp.readyState == 4 && xmlhttp.status == 200) {
18                      document.getElementById("Number").innerHTML =
19      xmlhttp.responseText;
20                  }
21              }
22          }
```

```
23        </script>
24    </head>
25    <body>
26    <div>
27          0-<input type = "text" value = "100" id = "maxValue" style = "width:2em"
28    />之间的随机数
29    <button type = "button" onclick = "GetNumber()">点击生成</button>
30    <div id = "Number"></div>
31    </div>
32    </body>
33    </html>
```

HomeController.cs 中的有关 Action：

```
1     public ActionResult S2()
2     {
3         return View();
4     }
5     public ActionResult GetNumber(int maxValue)
6     {
7         Random r = new Random();
8         int number = Convert.ToInt32(r.NextDouble() * maxValue);
9         return Content(number.ToString());
10    }
```

页面的运行结果如图9.3所示。单击"点击生成"按钮后，将在按钮下面出现如图9.4所示的随机数。

图9.3　页面的运行结果

图9.4　生成的随机数

注意：通过 Get 方式传递参数时，一般通过路由方式，如本例中的 maxValue。

【例9.3】通过Post方式传递参数。

S3.cshtml 中的代码：

```
1   @{
2       Layout = null;
3   }
4   <!DOCTYPE html>
5   <html>
6   <head>
7       <meta name="viewport" content="width=device-width" />
8       <title>S3</title>
9       <script type="text/javascript">
10          function GetNumber() {
11              var xmlhttp = new XMLHttpRequest();
12              var maxValue = document.getElementById("maxValue").value;
13              xmlhttp.open("Post", "/Home/GetNumber", true);
14              xmlhttp.setRequestHeader("Content-type", "application/x-www-
15      form-urlencoded");
16              xmlhttp.send("maxValue=" + maxValue);
17              xmlhttp.onreadystatechange = function () {
18                  if (xmlhttp.readyState == 4 && xmlhttp.status == 200) {
19                      document.getElementById("Number").innerHTML =
20      xmlhttp.responseText;
21                  }
22              }
23          }
24      </script>
25  </head>
26  <body>
27  <div>
28      0-<input type="text" value="100" id="maxValue" style="width:2em"
29  />之间的随机数
30  <button type="button" onclick="GetNumber()">点击生成</button>
31  <div id="Number"></div>
32  </div>
33  </body>
34  </html>
```

HomeController.cs 中的有关 Action：

```
1   public ActionResult S3()
2   {
3       return View();
4   }
5   public ActionResult GetNumber(int maxValue)
```

```
6       {
7           Random r = new Random();
8           int number = Convert.ToInt32(r.NextDouble() * maxValue);
9           return Content(number.ToString());
10      }
```

页面的运行结果同例 9.2 一致。本例通过 Post 的方式实现了例 9.2 的功能。

注意：在使用 Post 方式时，一般会传递参数。首先需要通过 setRequestHeader( )方法设定 HTTP 头，然后在 send( )方法中设定需要传递的参数。

9.2.3 获取响应

当服务器处理后，可以通过 XMLHttpRequest 对象的 responseText 属性来获取字符串形式的数据响应，或通过 responseXML 属性来获取 XML 形式的数据响应。例 9.1 ~ 例 9.3 均使用了 responseText 属性来获取响应。下面的例 9.4 将通过 responseXML 属性来获取响应。

【例 9.4】 通过 responseXML 属性获取数据响应。

S4.cshtml 中的代码：

```
1    @{
2        Layout = null;
3    }
4    <!DOCTYPE html>
5    <html>
6    <head>
7        <meta name="viewport" content="width=device-width" />
8        <title>S4</title>
9        <script type="text/javascript">
10           function GetStudent() {
11               var xmlhttp = new XMLHttpRequest();
12               xmlhttp.open("Get", "/Home/GetXml", true);
13               xmlhttp.send();
14               xmlhttp.onreadystatechange = function () {
15                   if (xmlhttp.readyState == 4 && xmlhttp.status == 200) {
16                       var xml = xmlhttp.responseXML;
17                       var text = "";
18                       var name = xml.getElementsByTagName("Name");
19                       var age = xml.getElementsByTagName("Age");
20                       var sex = xml.getElementsByTagName("Sex");
21                       for (i = 0; i < name.length; i++) {
22                           text = text + (i + 1) + ":姓名:" + name[i].childNodes
23   [0].nodeValue + " 年龄:" + age[i].childNodes[0].nodeValue + " 性别:" + sex
24   [i].childNodes[0].nodeValue + "<br/>";
25                       }
26                       document.getElementById("Students").innerHTML = text;
27                   }
```

```
28                  }
29              }
30      </script>
31    </head>
32    <body>
33      <div>
34        <button type = "button" onclick = "GetStudent()">获取学生信息</button>
35        <div id = "Students"></div>
36      </div>
37    </body>
38  </html>
```

HomeController.cs 中的有关 Action：

```
1   using System.Xml.Linq;
2   ...
3   public ActionResult S4()
4   {
5       return View();
6   }
7   public ActionResult GetXml()
8   {
9       XElement Datas = new XElement("Students");
10      List<string> name = new List<string>() { "张三", "李四", "王五" };
11      List<int> age = new List<int>() { 18, 30, 25 };
12      List<string> sex = new List<string>() { "男", "男", "女" };
13      for (int i = 0; i < 3; i++)
14      {
15          XElement Data = new XElement("Student");
16          XElement Name = new XElement("Name");
17          Name.SetValue(name[i]);
18          Data.Add(Name);
19          XElement Age = new XElement("Age");
20          Age.SetValue(age[i]);
21          Data.Add(Age);
22          XElement Sex = new XElement("Sex");
23          Sex.SetValue(sex[i]);
24          Data.Add(Sex);
25          Datas.Add(Data);
26      }
27      return Content(Datas.ToString(), "text/xml");
28  }
```

页面的运行结果如图 9.5 所示。单击"获取学生信息"按钮后，将在按钮下面出现如图 9.6 所示的学生信息。

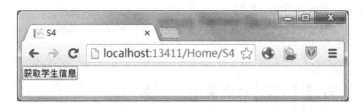

图9.5 页面的运行结果

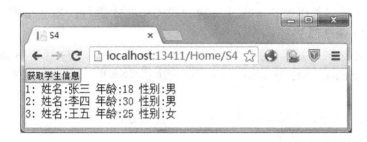

图9.6 获取的学生信息

注意： 在本例的控制器层中，使用了 XElement 对象来生成 XML 形式的数据，故需要在开始的位置添加 System. Xml. Linq 的引用。关于 XElement 对象的具体介绍，感兴趣的读者可以查看微软公司的官方说明。

9.2.4 onreadystatechange 事件

在获取服务器的响应时，需要知道服务器响应所处的状态，即 XMLHttpRequest 对象中的 readyState 属性。readyState 属性的取值范围为 0～4，分别对应请求未初始化、服务器链接已建立、请求已接收、请求处理中、请求已完成且响应已就绪这 5 种状态。

除此之外，还需要知道服务器响应是否正确地找到页面，即 XMLHttpRequest 对象中的 status 属性。status 属性的取值有 200 和 404 两种，分别对应正确找到和未找到页面。

XMLHttpRequest 对象中的 onreadystatechange 事件用于监视服务器的响应状况。当 readyState 属性为 4，status 属性为 200 时，表示响应已就绪，可以执行相应的页面操作。

在例 9.1～例 9.4 中，均通过 xmlhttp. onreadystatechange = function(){ } 来监视服务器的响应状况；通过 if (xmlhttp. readyState == 4 && xmlhttp. status == 200) 来判断响应是否就绪；最后，通过 if(){ } 中的子语句段，来执行服务器响应后的页面操作。

9.3 JQuery AJAX

虽然 XMLHttpRequest 对象已经能实现 AJAX，但是在使用上较为麻烦。而 JQuery 是一种比较优秀的 JavaScript 框架，能更为方便地处理 HTML 代码。因此，由 JQuery 提供的 AJAX 在使用上更受欢迎。本节将介绍 JQuery AJAX 中 3 种较为常用的实现方式。

9.3.1 load( )

根据 jquery. org 中的 API 文档说明，load( ) 方法用于从服务器获取数据，并将数据返回到匹配的选择器上。它的调用方式如下：

$(selector).load(url [, data] [, complete(responseText, textStatus, XMLHttpRequest)])

其中的选项说明如下。

selector：选择器，用于放置从服务器获取的数据。

url：必要，是获取服务器数据的地址。

data：可选，是向服务器传递的参数，可以是字符串（string）或者键-值（key：value）的集合。当 data 是字符串时，load( )方法通过 Get 方式进行；当 data 是键-值的集合时，load( )方法通过 Post 方式进行。

complete：可选，是回调函数（callback），在服务器响应后，页面更新前执行。responseText、textStatus、XMLHttpRequest 分别代表响应的数据、响应的状态、XMLHttpRequest 对象。

下面通过几个例子来说明 load( )方法的具体使用方式。

【例 9.5】load( )基本使用方式。

S5.cshtml 中的代码：

```
1    @{
2        Layout = null;
3    }
4    <!DOCTYPE html>
5    <html>
6    <head>
7        <meta name="viewport" content="width=device-width" />
8        <title>S5</title>
9        <script type="text/javascript" src="/Scripts/jquery-1.7.1.js">
10       </script>
11       <script type="text/javascript">
12           $(document).ready(function () {
13               $("#TimeBtn").click(function () {
14                   $("#Time").load("/Home/GetTime");
15               });
16           });
17       </script>
18   </head>
19   <body>
20       <div>
21           <input type="button" value="获取时间" id="TimeBtn" />
22           <div id="Time"></div>
23       </div>
24   </body>
25   </html>
```

HomeController.cs 中的有关 Action：

```
1    public ActionResult S5()
2    {
3        return View();
4    }
5    public ActionResult GetTime()
6    {
7        return Content(DateTime.Now.ToString());
8    }
```

本例通过 load( )方法实现了例 9.1 的功能,页面的运行效果同例 9.1。

【例 9.6】 参数的传递。

S6.cshtml 中的代码:

```
1    @{
2        Layout = null;
3    }
4    <!DOCTYPE html>
5    <html>
6    <head>
7    <meta name="viewport" content="width=device-width" />
8    <title>S6</title>
9    <script type="text/javascript" src="/Scripts/jquery-1.7.1.js">
10   </script>
11   <script type="text/javascript">
12       $(document).ready(function () {
13           $("#NumberBtn").click(function () {
14               $("#Number").load("/Home/GetNumber", "maxValue=100");
15           });
16           $("#NumberBtn2").click(function () {
17               $("#Number2").load("/Home/GetNumber", { maxValue: 100 });
18           });
19       });
20   </script>
21   </head>
22   <body>
23   <div>
24   <input type="button" value="get 方式" id="NumberBtn" />
25   <div id="Number"></div>
26   <br />
27   <input type="button" value="post 方式" id="NumberBtn2" />
28   <div id="Number2"></div>
29   </div>
30   </body>
31   </html>
```

HomeController.cs 中的有关 Action：

```
1      public ActionResult S6()
2      {
3          return View();
4      }
5      public ActionResult GetNumber(int maxValue)
6      {
7          Random r = new Random();
8          int number = Convert.ToInt32(r.NextDouble() * maxValue);
9          return Content(number.ToString());
10     }
11     [HttpPost]
12     public ActionResult GetNumber(string maxValue)
13     {
14         return Content(maxValue);
15     }
```

页面的运行结果如图 9.7 所示。分别单击"get 方式"按钮和"post 方式"按钮后，将在按钮下面分别出现如图 9.8 所示的数字。

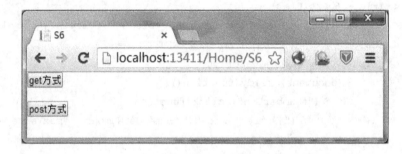

图 9.7　页面的运行结果

图 9.8　获取的数字

注意：因为"get 方式"按钮是通过字符串传递参数，所以通过 Get 方式实现，对应控制器层中的第一个 GetNumber() 方法，得到一个随机数；因为"post 方式"按钮通过键-值的集合传递参数，所以通过 Post 方式实现，对应控制器层中的第二个 GetNumber() 方法，得到一个固定数。

【例9.7】 回调函数的使用。

S7.cshtml 中的代码：

```
1   @{
2       Layout = null;
3   }
4   <!DOCTYPE html>
5   <html>
6   <head>
7       <meta name="viewport" content="width=device-width" />
8       <title>S7</title>
9       <script type="text/javascript" src="/Scripts/jquery-1.7.1.js">
10      </script>
11      <script type="text/javascript">
12          $(document).ready(function () {
13              $("#TimeBtn").click(function () {
14                  $("#Time").load("/Home/GetTime", function () {
15                      alert("服务器已响应!");
16                      $("#TimeTitle").html("服务器时间为:");
17                  });
18              });
19          });
20      </script>
21  </head>
22  <body>
23      <div>
24          <input type="button" value="获取时间" id="TimeBtn"/>
25          <div id="TimeTitle"></div>
26          <div id="Time"></div>
27      </div>
28  </body>
29  </html>
```

HomeController.cs 中的有关 Action：

```
1   public ActionResult S7()
2   {
3       return View();
4   }
5   public ActionResult GetTime()
6   {
7       return Content(DateTime.Now.ToString());
8   }
```

页面的运行结果如图 9.9 所示。单击"获取时间"按钮后，将先得到如图 9.10 所示的 JavaScript 提示框，"获取时间"按钮下面不刷新文字。单击"确定"按钮后，将在"获取时间"按钮下方出现如图 9.11 所示的文字和时间。

图 9.9　页面的运行结果

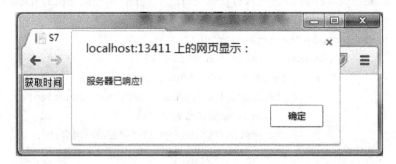

图 9.10　JavaScript 提示框

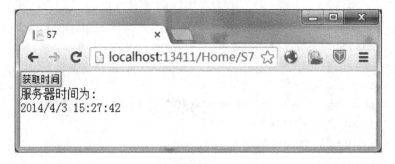

图 9.11　获得的文字和时间

注意： 由于回调函数在服务器响应后，页面更新前执行，因此，在用回调函数弹出 JavaScript 提示框时，不会先出现服务器的返回值，而会出现图 9.10 所示的结果。

9.3.2　get( )

根据 jquery.org 中的 API 文档说明，get( ) 方法用于通过 HTTP-Get 方式从服务器获取数据。它的调用方式如下：

```
$.get(url [, data ] [, success(data, textStatus, jqXHR) ] [, dataType ] )
```

其中的参数说明如下。

url：必要，是获取服务器数据的地址。

data：可选，是向服务器传递的参数，可以是字符串（string）或者键-值（key：value）的集合。

success：可选，是回调函数（callback）。在实现 AJAX 时，一般在回调函数中处理响应的数据，将其放在页面的合适位置。

dataType：可选，是期望获得的响应数据的格式。

下面通过几个例子来说明 get( )方法的具体使用方式。

【例 9.8】 get( )方法基本使用方式。

S8.cshtml 中的代码：

```
1    @{
2        Layout = null;
3    }
4    <!DOCTYPE html>
5    <html>
6    <head>
7    <meta name="viewport" content="width=device-width" />
8    <title>S8</title>
9    <script type="text/javascript" src="/Scripts/jquery-1.7.1.js">
10   </script>
11   <script type="text/javascript">
12       $(document).ready(function () {
13           $("#TimeBtn").click(function () {
14               $.get("/Home/GetTime", function (response) {
15                   $("#Time").html(response);
16               });
17           });
18       });
19   </script>
20   </head>
21   <body>
22   <div>
23   <input type="button" value="获取时间" id="TimeBtn" />
24   <div id="Time"></div>
25   </div>
26   </body>
27   </html>
```

HomeController.cs 中的有关 Action：

```
1    public ActionResult S8()
2    {
3        return View();
4    }
5    public ActionResult GetTime()
6    {
7        return Content(DateTime.Now.ToString());
8    }
```

本例通过 get() 方法实现了例 9.1 的功能，页面的运行效果同例 9.1。

注意：回调函数中的 response 是响应的数据。

【例 9.9】参数的传递。

S9.cshtml 中的代码：

```
1    @{
2        Layout = null;
3    }
4    <!DOCTYPE html>
5    <html>
6    <head>
7        <meta name="viewport" content="width=device-width" />
8        <title>S9</title>
9        <script type="text/javascript" src="/Scripts/jquery-1.7.1.js">
10       </script>
11       <script type="text/javascript">
12           $(document).ready(function () {
13               $("#NumberBtn").click(function () {
14                   $.get("/Home/GetNumber", "maxValue=100", function (response) {
15                       
16                       $("#Number").html(response);
17                   });
18               });
19               $("#NumberBtn2").click(function () {
20                   $.get("/Home/GetNumber", { maxValue: 100 }, function (response) {
21                       
22                       $("#Number2").html(response);
23                   });
24               });
25           });
26       </script>
27   </head>
28   <body>
29       <div>
30           <input type="button" value="字符串方式" id="NumberBtn" />
31           <div id="Number"></div>
32           <br/>
33           <input type="button" value="键-值的集合方式" id="NumberBtn2" />
34           <div id="Number2"></div>
35       </div>
36   </body>
37   </html>
```

HomeController.cs 中的有关 Action：

```
1      public ActionResult S9()
2      {
3          return View();
4      }
5      public ActionResult GetNumber(int maxValue)
6      {
7          Random r = new Random();
8          int number = Convert.ToInt32(r.NextDouble() * maxValue);
9          return Content(number.ToString());
10     }
11     [HttpPost]
12     public ActionResult GetNumber(string maxValue)
13     {
14         return Content(maxValue);
15     }
```

页面的运行结果如图 9.12 所示。分别单击"字符串方式"按钮和"键-值的集合方式"按钮后,将在按钮下面分别出现如图 9.13 所示的数字。

图 9.12　页面的运行结果

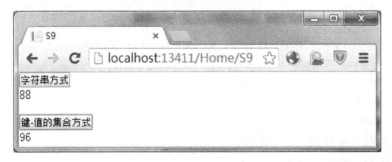

图 9.13　获取的数字

注意:和 load() 方法不同,get() 方法在传递参数时,无论是通过字符串方式还是键-值的集合方式,都将通过 Get 方式进行。因此,这两个按钮获得的数字都是随机数。

9.3.3　post()

根据 jquery.org 中的 API 文档说明,post() 方法用于通过 HTTP-Post 方式从服务器获取数据。它的调用方式如下：

$.post(url [, data] [, success(data, textStatus, jqXHR)] [, dataType])

因为各参数的含义和 get( )方法一致，故在此不再赘述。

下面通过几个例子来说明 post( )方法的具体使用方式。

【例 9.10】 post( )方法基本使用方式。

S10.cshtml 中的代码：

```
1     @{
2         Layout = null;
3     }
4     <!DOCTYPE html>
5     <html>
6     <head>
7         <meta name="viewport" content="width=device-width" />
8         <title>S10</title>
9         <script type="text/javascript" src="/Scripts/jquery-1.7.1.js">
10        </script>
11        <script type="text/javascript">
12            $(document).ready(function () {
13                $("#TimeBtn").click(function () {
14                    $.get("/Home/GetTime", function (response) {
15                        $("#Time").html(response);
16                    });
17                });
18            });
19        </script>
20    </head>
21    <body>
22        <div>
23            <input type="button" value="获取时间" id="TimeBtn"/>
24            <div id="Time"></div>
25        </div>
26    </body>
27    </html>
```

HomeController.cs 中的有关 Action：

```
1     public ActionResult S10()
2     {
3         return View();
4     }
5     public ActionResult GetTime()
6     {
7         return Content(DateTime.Now.ToString());
8     }
```

本例通过 post( )方法实现了例 9.1 的功能，页面的运行效果同例 9.1。

注意：回调函数中的 response 是响应的数据。

【例 9.11】 参数的传递。

S11.cshtml 中的代码：

```
1    @{
2        Layout = null;
3    }
4    <!DOCTYPE html>
5    <html>
6    <head>
7    <meta name="viewport" content="width=device-width" />
8    <title>S11</title>
9    <script type="text/javascript" src="/Scripts/jquery-1.7.1.js">
10   </script>
11   <script type="text/javascript">
12       $(document).ready(function () {
13           $("#NumberBtn").click(function () {
14               $.post("/Home/GetNumber", "maxValue=100", function (re
15   sponse) {
16                   $("#Number").html(response);
17               });
18           });
19           $("#NumberBtn2").click(function () {
20               $.post("/Home/GetNumber", { maxValue: 100 }, function (re
21   sponse) {
22                   $("#Number2").html(response);
23               });
24           });
25       });
26   </script>
27   </head>
28   <body>
29   <div>
30   <input type="button" value="字符串方式" id="NumberBtn"/>
31   <div id="Number"></div>
32   <br />
33   <input type="button" value="键-值的集合方式" id="NumberBtn2"/>
34   <div id="Number2"></div>
35   </div>
36   </body>
37   </html>
```

HomeController.cs 中的有关 Action：

```
1    public ActionResult S11()
2    {
3        return View();
4    }
5    public ActionResult GetNumber(int maxValue)
6    {
7        Random r = new Random();
8        int number = Convert.ToInt32(r.NextDouble() * maxValue);
9        return Content(number.ToString());
10   }
11   [HttpPost]
12   public ActionResult GetNumber(string maxValue)
13   {
14       return Content(maxValue);
15   }
```

页面的运行结果如图 9.14 所示。分别单击"字符串方式"按钮和"键-值的集合方式"按钮后，将在按钮下面分别出现如图 9.15 所示的数字。

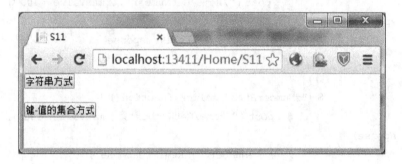

图 9.14 页面运行结果

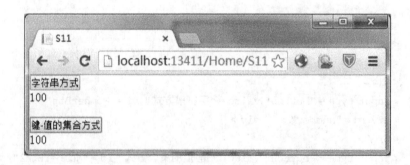

图 9.15 获取的数字

注意：和 load() 方法不同，post() 方法在传递参数时，无论是通过字符串方式，还是键-值的集合方式，都将通过 Post 方式进行。因此，这两个按钮获得的数字都是固定数。

9.4 综合实例

9.4.1 多属性查询

查询是系统中一个常用的功能。本节以 Student 表（数据表结构同例 2.1）为例，说明如何用 AJAX 方式实现多属性查询功能。

【例 9.12】查询功能。

(1) 视图层

查询功能的页面主要由两部分内容构成：一个是查询条件，本例中将 Student 表的所有属性都纳入查询条件中，进行多条件组合查询；另一个是查询结果，即符合查询条件的记录。为了区分这两块内容，本例在它们中间加了 <hr/>。页面的具体设计见 S12.cshtml。

S12.cshtml 中的代码：

```
1      @model IQueryable<S12.Models.Student>
2      @{
3          Layout = null;
4      }
5      <!DOCTYPE html>
6      <html>
7      <head>
8      <meta name="viewport" content="width=device-width" />
9      <title>S12</title>
10     <script type="text/javascript" src="/Scripts/jquery-1.7.1.js">
11     </script>
12     <script type="text/javascript">
13         $(document).ready(function () {
14             $("#submitBtn").click(function () {
15                 $.post("/Home/SearchStu", $("#paras").serialize(), function (response) {
16                     $("#Records").html(response);
17                 });
18             });
19         });
20     </script>
21     </head>
22     <body>
23     <div style="padding: 1px; margin: 10px; font-family: Arial, sans-serif;
24     font-size: small;">
25     <form id="paras">
26     学号:<input type="text" style="width: 50px" name="sno" />
27     姓名:<input type="text" style="width: 50px" name="sname" />
```

```
29      性别:<select name="ssex"><option value=""></option>
30      <option value="男">男</option>
31      <option value="女">女</option>
32      </select>
33      年龄:<input type="text" style="width:50px" name="sage" />
34      所在系所:<input type="text" style="width:50px" name="sdept" />
35      <input type="button" value="查询" id="submitBtn" />
36      </form>
37      </div>
38      <hr style="width:96%" />
39      <div id="Records">
40          @Html.Partial("StudentRecord", Model)
41      </div>
42      </body>
43      </html>
```

本例中使用 JQuery 中的 post() 方法实现了 AJAX 查询。由于查询结果部分的页面设计会被反复调用,因此将该部分内容放置到分部页中,即 StudentRecord.cshtml。为了使查询结果看上去更友好,本例中使用了少量的 CSS 对页面布局进行了美化。

StudentRecord.cshtml 中的代码:

```
1       @model IQueryable<S12.Models.Student>
2       <style type="text/css">
3           table
4           {
5               padding: 1px;
6               margin: 10px;
7               font-family: Arial, sans-serif;
8               font-size: small;
9               table-layout: auto;
10              border-collapse: collapse;
11              border: 1px solid #C0C0C0;
12              width: 96%;
13          }
14          td
15          {
16              padding: 1px;
17              margin: 1px;
18              border: 1px solid #C0C0C0;
19              text-align: center;
20          }
21      </style>
22      <table>
23      <tr>
```

```
24          <td>学号</td>
25          <td>姓名</td>
26          <td>性别</td>
27          <td>年龄</td>
28          <td>所在系所</td>
29        </tr>
30        @{
31            foreach (var student in Model)
32            {
33          <tr>
34          <td>@student.Sno</td>
35          <td>@student.Sname</td>
36          <td>@student.Ssex</td>
37          <td>@student.Sage</td>
38          <td>@student.Sdept</td>
39          </tr>
40            }
41        }
42        </table>
```

（2）模型层

为了从数据库获取数据，本例首先建立了数据连接文件 Student.edmx（具体建立过程可参考第 2 章）。本例中，模型层的主要任务是根据查询条件，从数据库获取查询结果。由于查询条件个数的不确定性，因此本例中的输入参数设定为强类型 Student；之后，通过判断属性是否为空，来确定是否进行相应属性的查询。具体的处理过程见 StudentRep.cs 中的 GetStudentsByKeys() 方法。

在页面初始化时，设定页面显示所有学生信息。因此，需要一个获取全体学生信息的方法，即 StudentRep.cs 中的 GetStudents()。

StudentRep.cs 中的代码：

```
1     using System;
2     using System.Collections.Generic;
3     using System.Linq;
4     using System.Web;
5     namespace S12.Models
6     {
7         public class StudentRep
8         {
9             StudentEntities db = new StudentEntities();
10            public IQueryable<Student> GetStudents()
11            {
12                return db.Students;
13            }
14            public IQueryable<Student> GetStudentsByKeys(Student s)
```

```
15      {
16          IQueryable<Student> students = db.Students;
17          if (!string.IsNullOrWhiteSpace(s.Sno))
18              students = students.Where(q => q.Sno == s.Sno);
19          if (!string.IsNullOrWhiteSpace(s.Sname))
20              students = students.Where(q => q.Sname.Contains(s.Sname));
21          if (!string.IsNullOrWhiteSpace(s.Ssex))
22              students = students.Where(q => q.Ssex == s.Ssex);
23          if (s.Sage != null)
24              students = students.Where(q => q.Sage == s.Sage);
25          if (!string.IsNullOrWhiteSpace(s.Sdept))
26              students = students.Where(q => q.Sdept.Contains(s.Sdept));
27          return students;
28      }
29  }
30 }
```

(3) 控制器层

由于本例中涉及模型层的操作，因此在顶部先添加相应文件的引用，即"using S12.Models;"；同时进行相应类的实例化，即"StudentRep StuRep = new StudentRep();"。

由于模型层中需要的输入参数是强类型 Student，因此在控制器层需要把表单中的参数转换成 Student 中的对应属性，具体过程见 SearchStu() 方法。

HomeController.cs 中的有关 Action：

```
1   using S12.Models;
2   …
3   StudentRep StuRep = new StudentRep();
4   public ActionResult S12()
5   {
6       IQueryable<Student> students = StuRep.GetStudents();
7       return View(students);
8   }
9   [HttpPost]
10  public ActionResult SearchStu(FormCollection collection)
11  {
12      Student stu = new Student();
13      stu.Sno = collection["sno"];
14      stu.Sname = collection["sname"];
15      stu.Ssex = collection["ssex"];
16      try
17      {
18          stu.Sage = Convert.ToDecimal(collection["sage"]);
19      }
20      catch
```

```
21        {
22            stu.Sage = null;
23        }
24        stu.Sdept = collection["sdept"];
25        IQueryable<Student> students = StuRep.GetStudentsByKeys(stu);
26        return PartialView("StudentRecord", students);
27    }
```

（4）运行结果

页面的初始运行结果如图 9.16 所示。"性别"选择"男","年龄"文本框中输入"19",单击"查询"按钮后,得到如图 9.17 所示的查询结果。由于在模型层中,关于姓名和所在系所的查询使用了 Contains() 方法,因此,这两个字段可以进行模糊查询。在"姓名"文本框中输入"张",单击"查询"按钮后,得到如图 9.18 所示的查询结果。

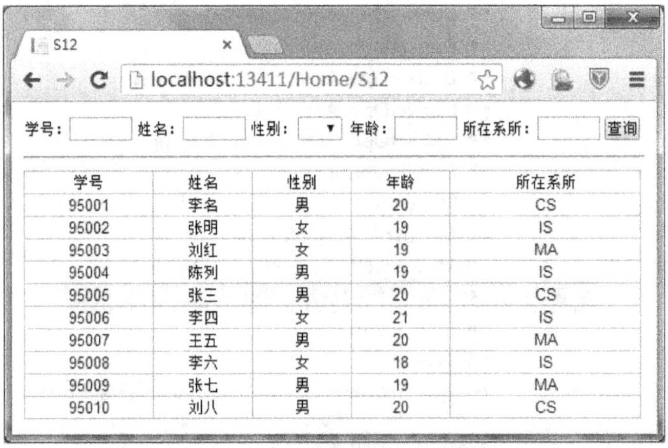

图 9.16　页面的初始运行结果

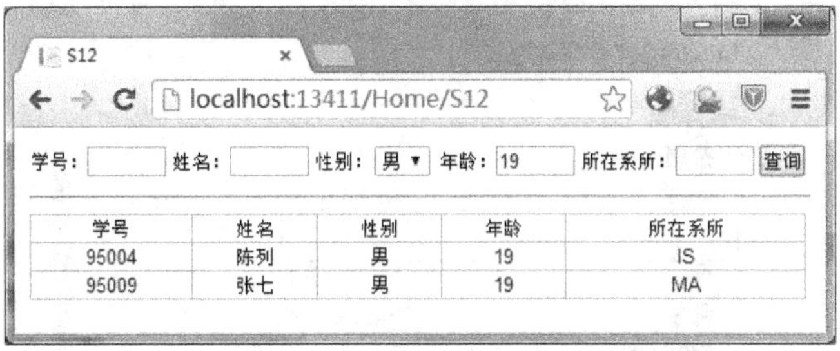

图 9.17　多属性查询

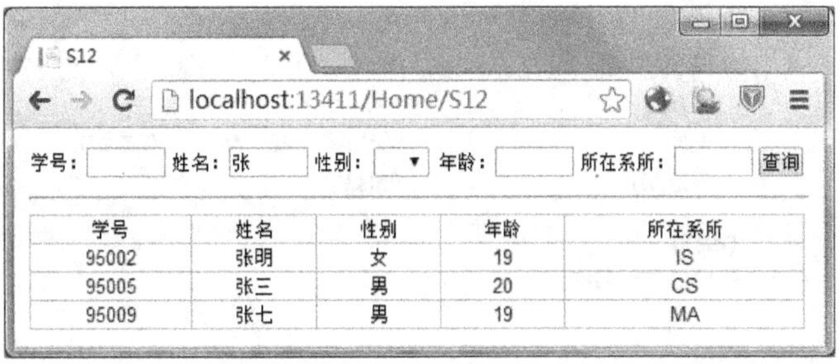

图 9.18　模糊查询

9.4.2 分页显示

当数据记录较多时，一般需要分页显示。例 9.13 在例 9.12 的基础上，通过 MvcPager 控件 V2.0（杨涛，2014a），用 AJAX 方式实现了分页显示的功能。

【例 9.13】分页显示功能。

(1) 前置步骤

需要先下载 MvcPager.dll，然后在引用中添加该文件，如图 9.19 所示。（建议选择版本号高于 2.0 的版本，早期版本可能不兼容 MVC 4。）

图 9.19　引用 MvcPager.dll

（2）视图层

由于使用了第三方控件，因此视图文件的开头要添加控件的引用，即"@ using Webdiyer. WebControls. Mvc;"。同时视图文件的强类型变为 PagedList < S13. Models. Student >。

S13. cshtml 中的代码：

```
1    @ using Webdiyer. WebControls. Mvc;
2    @ model PagedList < S13. Models. Student >
3    @ {
4        Layout = null;
5    }
6    <! DOCTYPE html >
7    < html >
8    < head >
9    < meta name = "viewport" content = "width = device-width" />
10   < title > S13 </ title >
11   < script type = " text/javascript " src = "/Scripts/jquery-1. 7. 1. js " >
12   </ script >
13   < script type = "text/javascript" >
14       $ (document). ready(function () {
15           $ ("#submitBtn"). click (function () {
16               $ . post ("/Home/SearchStus", $ ("#paras"). serialize (), func-
17   tion (response) {
18                   $ ("#Records"). html (response);
19               });
20           });
21       });
22   </ script >
23   </ head >
24   < body >
25   < div style = " padding: 1px; margin: 10px; font-family: Arial, sans-serif;
26   font-size: small;" >
27   < form id = "paras" >
28   学号: < input type = "text" style = "width: 50px" name = "sno" />
29   姓名: < input type = "text" style = "width: 50px" name = "sname" />
30   性别: < select name = "ssex" > < option value = "" > </ option >
31   < option value = "男" >男 </ option >
32   < option value = "女" >女 </ option >
33   </ select >
34   年龄: < input type = "text" style = "width: 50px" name = "sage" />
35   所在系所: < input type = "text" style = "width: 50px" name = "sdept" />
36   < input type = "button" value = "查询" id = "submitBtn" />
37   </ form >
38   </ div >
```

```
39      <hr style = "width: 96%" />
40      <div id = "Records">
41          @Html.Partial("Record", Model)
42      </div>
43      </body>
44      </html>
```

在分部页 Record.cshtml 中，需要使用添加分页条的方法@Ajax.Pager()。有关该方法的具体说明，感兴趣的读者可以查看 MvcPager 控件的帮助文档（杨涛，2014b）。同时，需要注册 AJAX 文件，即@{Html.RegisterMvcPagerScriptResource();}。若不注册该文件，则翻页将以普通的方式进行。

Record.cshtml 中的代码：

```
1       @using Webdiyer.WebControls.Mvc;
2       @model PagedList<S13.Models.Student>
3       <style type = "text/css">
4           table
5           {
6               padding: 1px;
7               margin: 10px;
8               font-family: Arial, sans-serif;
9               font-size: small;
10              table-layout: auto;
11              border-collapse: collapse;
12              border: 1px solid #C0C0C0;
13              width: 96%;
14          }
15          td
16          {
17              padding: 1px;
18              margin: 1px;
19              border: 1px solid #C0C0C0;
20              text-align: center;
21          }
22      </style>
23      <table>
24      <tr>
25      <td>学号</td>
26      <td>姓名</td>
27      <td>性别</td>
28      <td>年龄</td>
29      <td>所在系所</td>
30      </tr>
31          @{
```

```
32              foreach (var student in Model)
33              {
34      <tr>
35      <td>@student.Sno</td>
36      <td>@student.Sname</td>
37      <td>@student.Ssex</td>
38      <td>@student.Sage</td>
39      <td>@student.Sdept</td>
40      </tr>
41              }
42          }
43      </table>
44      <div style = "padding: 1px; margin: 10px; font-family: Arial, sans-serif;
45      font-size: small; text-align: center">
46          @Ajax.Pager(Model, "Default", new { key = ViewData["key"] }, new PagerOp
47      tions { PageIndexParameterName = "pid" }, new MvcAjaxOptions { UpdateTargetId
48      = "Records", HttpMethod = "Get" }, new { })
49          @{Html.RegisterMvcPagerScriptResource();}
50      </div>
```

(3) 模型层

模型层的内容无变化，同例9.12。

(4) 控制器层

在例9.12的基础上，本例中控制器层需要完成额外的3项操作：

1) 查询条件的持续传递。在进行查询操作后，进行翻页操作时，需要用查询条件去限定翻页的原始数据；否则，分页结果将与查询条件不符，引起用户误解。本例中，在查询操作后，先将查询条件拼装成1个字符串（Post方式的SearchStus()方法中的key），然后传递给视图层（Record.cshtml），放入翻页的超链接中。在单击超链接翻页时，再对字符串进行拆分，获取相应参数（Get方式的SearchStus()方法中的处理方法）。

2) 统一排序。为了保证翻页前后数据的一致性，需要用同一种方法对从模型层获取的数据进行排序。本例中，统一对学号Sno进行了升序排列，即OrderBy(s = >s.Sno)。

3) 提取翻页后的数据。使用ToPagedList（pageIndex，pageSize）方法从排序后的数据中提取第pageIndex项到第pageIndex + pageSize项。

HomeController.cs中的有关Action：

```
1       using Webdiyer.WebControls.Mvc;
2       …
3       public ActionResult S13(int pid)
4       {
5           if (pid == null)//首次载入
6           {
7               PagedList<Student> students = StuRep.GetStudents().OrderBy(s = >
8       s.Sno).ToPagedList(1,5);
```

```csharp
9            return View(students);
10        }
11        else//翻页后载入
12        {
13            PagedList<Student> students = StuRep.GetStudents().OrderBy(s =>
14  s.Sno).ToPagedList(pid 1, 5);
15            return PartialView("Record",students);
16        }
17    }
18    [HttpPost]
19    public ActionResult SearchStus(FormCollection collection)
20    {
21        Student stu = new Student();
22        string key = "";
23        stu.Sno = collection["sno"];
24        key = key + collection["sno"] + ",";
25        stu.Sname = collection["sname"];
26        key = key + collection["sname"] + ",";
27        stu.Ssex = collection["ssex"];
28        key = key + collection["ssex"] + ",";
29        try
30        {
31            stu.Sage = Convert.ToDecimal(collection["sage"]);
32        }
33        catch
34        {
35            stu.Sage = null;
36        }
37        key = key + collection["sage"] + ",";
38        stu.Sdept = collection["sdept"];
39        key = key + collection["sdept"];
40        PagedList<Student> students = StuRep.GetStudentsByKeys(stu).OrderBy(s
41  => s.Sno).ToPagedList(1,5);
42        ViewData["key"] = key;
43        return PartialView("Record", students);
44    }
45    [HttpGet]
46    public ActionResult SearchStus(string key, int pid)
47    {
48        ViewData["key"] = key;
49        string[] keyValue = key.Split(new char[] { ',' });
50        Student stu = new Student();
51        stu.Sno = keyValue[0];
```

```
52          stu.Sname = keyValue[1];
53          stu.Ssex = keyValue[2];
54          try
55          {
56              stu.Sage = Convert.ToDecimal(keyValue[3]);
57          }
58          catch
59          {
60              stu.Sage = null;
61          }
62          stu.Sdept = keyValue[4];
63          PagedList<Student> students = StuRep.GetStudentsByKeys(stu).OrderBy(s
64       => s.Sno).ToPagedList(pid 1, 5);
65          return PartialView("Record", students);
66      }
```

(5) 运行结果

页面的初始运行结果如图9.20所示。直接单击第2页，将得到如图9.21所示的翻页结果。"性别"选择"男"，单击"查询"按钮后，将得到如图9.22所示的查询结果。此时，再单击第2页，将得到如图9.23所示的翻页后的查询结果。

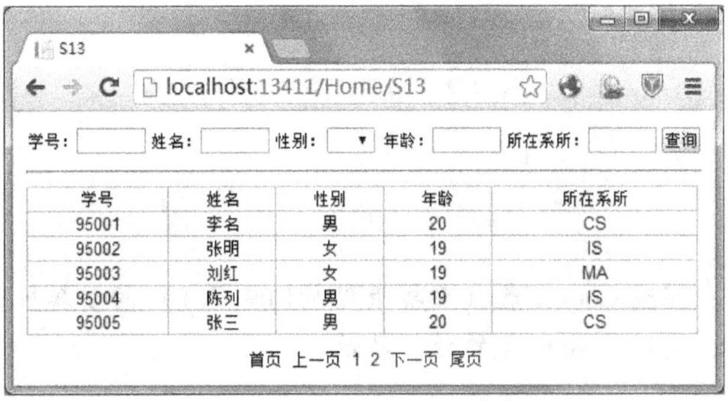

图9.20　页面的初始运行结果

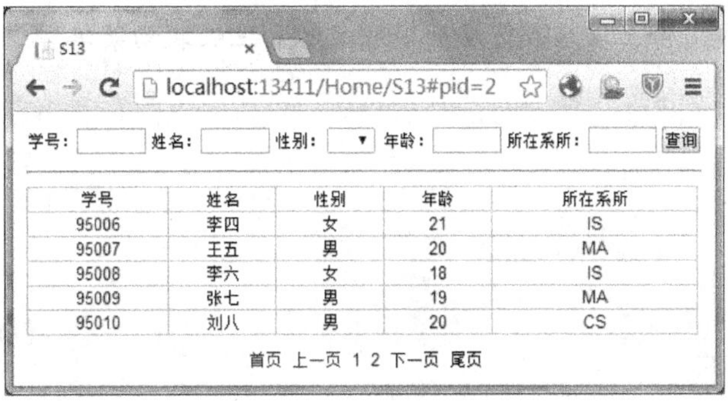

图9.21　直接翻页

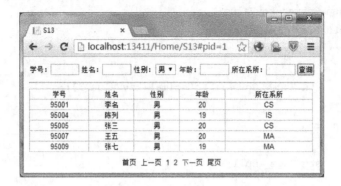

图 9.22 查询结果

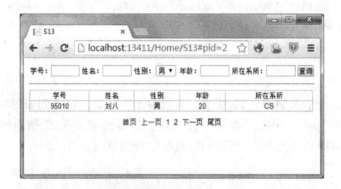

图 9.23 翻页后的查询结果

※习　题

1. 仿照例 9.12，实现 Course 表（数据表结构同例 2.1）的多属性查询。
2. 仿照例 9.13，实现 Course 表的分页显示。

※综合应用

使用 AJAX 完成省份、城市和区域的地理名称的三级联动菜单。

第 10 章

服务器（IIS）的配置与使用

服务器部署是软件工程项目的最后一个环节，也是不可或缺的一个环节。由于本书使用微软公司的 ASP.NET MVC 4 框架，因此服务器选用微软公司的 IIS 服务器。本章选用 IIS 7 作为部署讲解的对象，通过逐步讲解 IIS 的安装、属性与配置以及工程的发布，使读者了解整个软件发布的流程。

10.1 IIS 简介

IIS 的全称为 Internet Information Services，即 Internet 信息服务，是微软公司提供的在 Windows 上运行的互联网基本服务。Windows 7 下的 IIS 版本为 7.0，提供"FTP 服务器"、"Web 管理工具"和"万维网服务"三项内容。刚刚接触它的读者，可以简单地将 IIS 理解为在 Windows 系统下发布网站的工具。

10.2 IIS 安装

笔者使用的操作系统是 Windows 7，家庭普通版即可安装使用 IIS，但是不能配置和使用全部的功能，专业版及以上版本可以使用 IIS 的全部功能。下面将描述如何在 Windows 7 中安装 IIS。

Step01：打开"控制面板"，找到"程序"，单击进入，如图 10.1 所示。

图 10.1 控制面板

Step02：选择"打开或关闭 Windows 功能"，如图 10.2 所示。

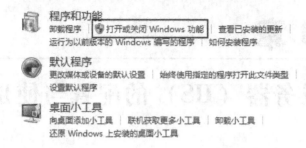

图 10.2　打开或关闭 Windows 功能

Step03：将"Internet 信息服务"一项根据需要进行选择性的安装，如图 10.3 所示。

图 10.3　选择 Internet 信息服务

Step04：单击"确定"按钮后，系统开始安装选定的功能，等待几分钟后，IIS 安装完毕，如图 10.4 所示。

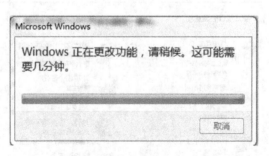

图 10.4　等待安装

Step05：右击"计算机"，在弹出的快捷菜单中选择"管理"，会出现"计算机管理"的内容。这时会发现较平时多了"Internet 信息服务（IIS）管理器"一项，如图 10.5 所示。

单击该项就会出现 IIS 管理器的界面，如图 10.6 所示。

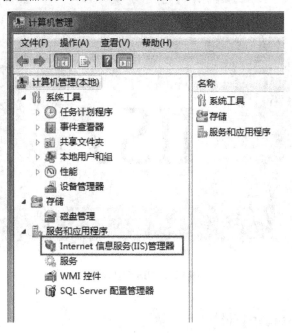

图 10.5　计算机管理

图 10.6　IIS 管理器

Step06：刚刚安装好的 IIS 提供一个自带的网站，单击浏览后会打开如图 10.7 所示的网页。如果网页正确显示，则说明 IIS 已经安装成功。

图 10.7 IIS 默认网站

10.3 IIS 的属性与配置

IIS 7.0 提供了多种属性可以供用户进行选择与设置，如配置"连接字符串"、"日志"功能等。单击 IIS 默认网站后就可以看到所有的内容，如图 10.8 所示。

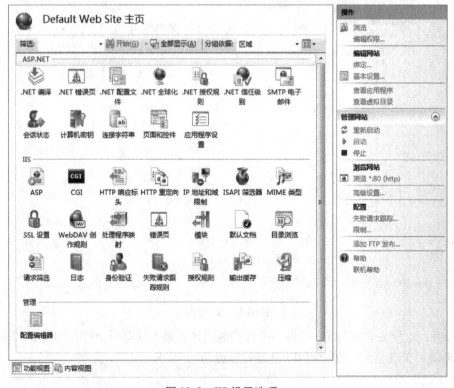

图 10.8 IIS 设置选项

虽然已经安装了 Visual Studio 2012，但是由于刚刚安装好 IIS，因此应用程序池只有 .NET Framework 2.0 的版本，如图 10.9 所示。

图 10.9　应用程序池

发布 MVC 4.0 的网站，需要 .NET Framework 4.0 的应用程序池。因为已经安装了 Visual Studio 2012，所以我们只需要注册一下即可。

首先，运行 CMD 程序。可以使用 <Win + R> 快捷键调出"运行"，然后输入"cmd"打开该程序，如图 10.10 所示。

图 10.10　运行 CMD 程序

然后，输入"C：\Windows\Microsoft.NET\Framework\v4.0.30319\aspnet_regiis.exe-i"，按回车键运行，如图 10.11 所示。

图 10.11　进行注册

等待片刻后即可安装成功，如图 10.12 所示。

图 10.12 安装成功

这时，再去看应用程序池的选项，已经出现 NET Framework 4.0 的选项，如图 10.13 所示。

图 10.13 应用程序池

至此，关于 IIS 的配置已经全部完成，现在就只有等待网站的完成，并进行发布了。

10.4 工程在 IIS 上的发布

一个网站做好后就可以发布了。IIS 7.0 已经非常智能了，发布一个网站时并不需要烦琐的过程与设置，只需要几个步骤就可以把做好的网站发布出来。具体的发布步骤如下：

Step01：首先，需要更改 MVC 4 程序中的设置，单击"解决方案资源管理器"中的"引用"，找到图 10.14 所示的三项引用的内容。

Step02：右击"System.Web.Abstractions"，在弹出的快捷菜单中选择"属性"，更改"属性"中的"复制本地"，将其值更改为"True"，如图 10.15 所示。值得注意的是，这 3 个引用的"复制本地"都需要修改。

Step03：然后打开 IIS 管理器（右击"计算机"，然后在弹出的快捷菜单中选择"管理"），选择"添加网站"一项，如图 10.16 所示。也可以通过右键添加。

第 10 章 服务器（IIS）的配置与使用

图 10.14 需要更改的引用

图 10.15 更改内容

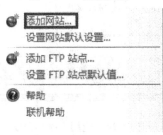

图 10.16 添加网站

Step04：此时会出现如图 10.17 所示的对话框，我们需要填写"网站名称"，选择"应用程序池"和"物理路径"，以及设定绑定的"IP 地址"和"端口"。因为没有设置 DNS 服务器，所以不用填写"主机名"。

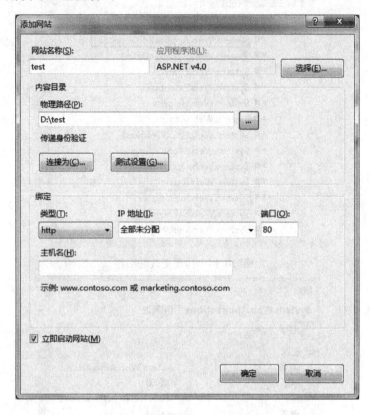

图 10.17 "添加网站"对话框

Step05：应用程序池选择"ASP.NET v4.0"一项，如图 10.18 所示。

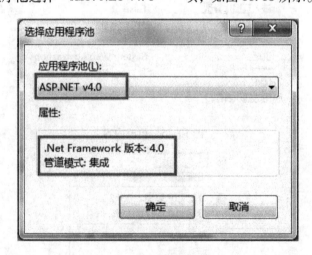

图 10.18 应用程序池选项

一般情况下，"IP 地址"一项默认即可，"端口"一项理论上只要不被其他服务占用即可，笔者比较推荐的端口有"80"、"8080"和"8087"。

配置完后，单击"确定"按钮即可发布网站，可以在右侧的"网站管理"中选择浏览网站，也可以进行其他对网站的管理操作，如图 10.19 所示。在浏览器中输入 http：//localhost：80 也可访问发布的网站。

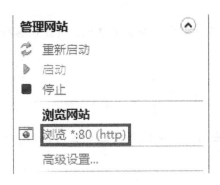

图 10.19　网站管理

※习　题

1. 新建一个 MVC 4 工程，将工程发布到 IIS 7 上。要求：
1）端口号分别为 80 和 8090。
2）修改网址设置，使得在 8090 端口下不需要输入端口号即可直接进入页面。
2. 设置页面防火墙为开启，将工程添加到防火墙中。
3. 在 IIS 中修改默认页面：在默认工程中，默认的显示页面是 Home 页面，需在 IIS 中修改配置，将默认页面修改为 Register 页面。
4. 当网站发出 HTTP 500 错误时，在 IIS 中应该如何解决？
一般来说问题在于：
1）系统可能没有注册 msjetoledb40.dll。
解决办法：点"开始→运行"，输入 regsvr32 msjetoledb40.dll，单击回车即可；
2）数据库所在文件夹权限没有打开。
解决办法：打开计算机，然后点菜单上的"文件夹选项"→"查看"，然后把"使用简单文件夹共享（推荐）"前面的钩去掉，然后单击"确认"；接下来回到需要打开权限的文件夹（即数据库存放的文件夹），进行操作："右键"→"属性"→"安全"，单击"添加"→"高级"→"立即查找"，然后在下面的地方选择"everyone"→单击"确定"→单击刚才加入的"everyone"，然后在下面的大框编辑 everyone 权限为完全控制，保存即可。
3）可能存在需要打开 guest 用户的问题。
解决办法："计算机"→"控制面板"→"管理工具"→"计算机管理"→"本地用户和组"→"用户"→"guest"，双击"guest"，然后把弹出框上的"账户已停用"前的钩去掉，保存即可。
4）temp（临时文件夹）权限需要打开。
解决办法："C 盘"→"windows"→"temp"，单击右键→"属性"，就会看到一个叫做"安全"的选项，添加一个 everyone，权限设置为完全控制，再将你正在使用 windows 的用户也设置为完全控制。（此修改步骤类似问题2，可以参照修改）

另外：主要是 IIS 服务器没有开启父路径，修改方法：

在 IIS 中的"属性"→"主目录"→"配置"→"选项"，在"启用父路径"前面打上勾。确认刷新。

如果服务器提供商出于安全考虑不开启父路径，则建议路径指向的时候写绝对路径，即完整地址。

※综合应用

将第2、3和5章综合应用中完成的实例进行应用发布。发布的端口设置为默认的80端口。

参考文献

[1] 马颖华，苏贵洋，袁艺，苏桂涛. ASP.NET 2.0 网络编程从基础到实践［M］. 北京：电子工业出版社，2007.

[2] Anytao. 使用 ActionSelector 控制 Action 的选择［J/OL］. 博客园［2009-4-21］. http://www.cnblogs.com/anytao/archive/2009/04/22/anytao-mvc-01-actioninrole.html.

[3] Johnny Yan. MVC 自定义 AuthorizeAttribute 实现权限管理［J/OL］. 博客园［2012-7-24］. http://www.cnblogs.com/jyan/archive/2012/07/24/2606646.html.

[4] 苏飞. 请慎用 ASP.Net 的 validateRequest = "false" 属性［J/OL］. 博客园［2009-5-16］. http://www.cnblogs.com/sufei/archive/2009/05/16/1485980.html.

[5] sslyc8991. 缓存，弹出提示［J/OL］. CSDN 博客［2013-9-2］. http://blog.csdn.net/sslyc8991/article/details/10916633.

[6] 无恨星辰. MVC 多级 Views 目录 asp.net mvc4 路由重写及 修改 view 的寻找视图的规则［J/OL］. 博客园［2013-9-17］. http://www.cnblogs.com/weixing/p/3326188.html.

[7] 懒人之家. jquery 制作仿 iPhone 苹果手机界面触屏切换效果［J/OL］. 懒人之家［2012-7-5］. http://www.lanrenzhijia.com/jquery/460.html.

[8] 17 素材. jquery 图片放大镜插件制作多种图片放大查看效果［J/OL］. 17 素材［2013-5-6］. http://www.17sucai.com/pins/358.html.

[9] Raul Iloc. MVC 4 网站中集成 jqGrid 表格插件［J/OL］. 滴答的雨，译. CSDN 博客［2014-4-15］. http://www.cnblogs.com/heyuquan/p/3665286.html.

[10] Jesse James Garrett. Ajax: A New Approach to Web Applications［J/OL］. adaptive path［2005-2-18］. http://www.adaptivepath.com/ideas/ajax-new-approach-web-applications/.

[11] 杨涛. MVCPager 概述［J/OL］. 杨涛主页. http://www.webdiyer.com/mvcpager.

[12] 杨涛. MVCPager 帮助文档 — AjaxHelper.Pager 扩展方法［J/OL］. 杨涛主页. http://www.webdiyer.com/mvcpager/docs/ajaxpager.